HIDDEN
REALMS

A celebration of 100 of the
finest caves and mines in
Great Britain and Ireland

HIDDEN REALMS

MARTYN FARR

Vertebrate Publishing, Sheffield
www.adventurebooks.com

A celebration of 100 of the
finest caves and mines in
Great Britain and Ireland

HIDDEN
REALMS

MARTYN FARR

First published in 2023 by Vertebrate Publishing.

 VERTEBRATE PUBLISHING
Omega Court, 352 Cemetery Road, Sheffield S11 8FT, United Kingdom.
www.adventurebooks.com

Copyright © 2023 Martyn Farr and Vertebrate Publishing Ltd.

Front cover: Powell's Lode Cavern.
Frontispiece: Piccadilly in Ogof Ffynnon Ddu 2.

Photography by Martyn Farr.

Martyn Farr has asserted his rights under the Copyright, Designs and Patents Act 1988 to be identified as author of this work.

A CIP catalogue record for this book is available from the British Library.

ISBN 978-1-83981-081-7

All rights reserved. No part of this work covered by the copyright herein may be reproduced or used in any form or by any means – graphic, electronic, or mechanised, including photocopying, recording, taping, or information storage and retrieval systems – without the written permission of the publisher.

Edited by Helen Parry; design, maps and production by Jane Beagley.
www.adventurebooks.com

Vertebrate Publishing is committed to printing on paper from sustainable sources.

Printed and bound in China by Latitude Press.

Caving is an activity that carries a risk of personal injury or death. Participants must be aware of and accept that these risks are present and they should be responsible for their own actions and involvement. Nobody involved in the writing and production of this book accepts any responsibility for any errors that it may contain, nor are they liable for any injuries or damage that may arise from its use. All caving is inherently dangerous and the fact that individual descriptions in this volume do not point out such dangers does not mean that they do not exist. Take care.

Always research access for indvidual caves before you go, as access arrangements and the legality of entering a particular site can change over time.

Contents

PREFACE vi
INTRODUCTION vi
ACKNOWLEDGEMENTS vii

South Wales 1
Mid and North Wales 43
Mendips and Southern England 65
Forest of Dean 101
Peak District and Central England 109
Yorkshire and Northern England 131
Scotland 179
Ireland 187

GLOSSARY 216

Preface

All of the caves and mines in this book are ones that I have personally visited and enjoyed over many years. I was privileged to be part of the discovery, early exploration or extension of a number of the sites and much of this I recorded photographically at the time. In compiling *Hidden Realms* I felt it was important to revisit as many of these sites as possible, and so, for the past two and a half years, this has been my mission. This enables me to present you, the reader, with many modern, quality images, complemented by some historical pictures from my library. I hope you enjoy this offering.

Introduction

This is the book I have always dreamed of – a photographic celebration of the underworld across the British Isles. Each site is portrayed through images that I hope will captivate and inspire you, and the few words I have written are there simply to add a little context.

I have been an underground explorer and enthusiast for over sixty years. It has been my lifelong passion and the elemental magic that this environment evokes has never waned. Yes, I understand that few people will be so fanatical, but for me there is something in this subterranean world that is as obsessive as a drug. Very simply, I love the physical challenge, the intense thrill of exploration, the camaraderie; and it all blends into one in a world of often amazing, wonderful, strange and beautiful sights. From ten years of age I was hooked, and, over time, photography became a complementary interest. It is one of the tools that enables me to show people what is down there, to share some of the wonders and to help explain why I love doing it. An image speaks a thousand words. Sixty years later the attraction is as strong as ever, and thankfully I am still fit enough to participate at a reasonable level. This is my opportunity to share my enthusiasm with you.

The sites included in this book have been chosen for a myriad of reasons; some for their size and splendour, some for sporting challenge, some for historical or archaeological interest, and yet others for their long-standing, popular appeal. Not all involve difficult caving and indeed some, in part, are show caves or mines accessible to all. Of the thousands of sites across the British Isles, one of the biggest dilemmas was which caves, or mines, to leave out. I apologise to those of you who think your favourite venue should have been included, but at the end of the day there were extremely difficult decisions to be made.

So, here is a window into my world, a snapshot of some of the places I have visited and some of the interesting and fabulous scenes I have captured. A number of the images are old and thus not presented at as high a standard as modern imaging techniques allow. However, I am sure that they will inspire nostalgia for many and to me they convey something special; an evocative memory or a historic exploratory event.

This book targets a wide audience and presents just a sample of the esoteric world beneath our feet. Many of you will want to see some of these places for yourself and I wholeheartedly endorse that desire. You don't have to be rich or highly educated to pursue this sport (I certainly was not!). Be aware that *Hidden Realms* is not a guidebook and it was never conceived in that light. You will need to research (thoroughly) and work out for yourself how best to visit these places. Some sites are easy to access, some are difficult and some currently have no officially permitted access (I was lucky enough to be granted access to a couple of sites for this book which are not normally accessible).

If you choose to seek out some of these wonderful locations you will need to exercise care – care not to occasion a rescue and, equally importantly, care for and conservation of the underground environment. Approaching your local caving club or mining society would be a responsible place to start. Remember: for the ill-informed, poorly equipped and perhaps foolhardy, these places present hazards which are potentially life-threatening. Ultimately, your safety is your responsibility.

Hidden Realms is a celebration of a unique and wonderful heritage. At whatever level you choose to explore this magical underworld, I guarantee you will not be disappointed.

Acknowledgements

While all the images in this book were taken by me, it is essential that I recognise the immense help provided by my companions at each of the sites visited. Every image is a snapshot in time but none of them could have been successfully captured without the other fantastically supportive people involved. There are those who facilitated a visit by supplying practical advice or arranging access, those who organised helpers and those who took on the role of trip leader. There were places to rig, equipment to be carried and lengthy periods standing around – all in cold, often wet and draughty conditions. To all of the people who have been involved in the creation of this book, I am eternally indebted.

I would like to recognise the particular dedication of a few helpers who each gave up their time on numerous occasions. In this respect I would mention Rodney Beaumont, Kevin Gannon, Martin Grass, Pat Cronin, Linda Windham, Dave McDonogh, Adele Ward, Greg Brock, Hugh Norton, Louise McMahon, Phil Rowsell and Paul 'Beardy' Swire.

I also thank the countless landowners, the show cave and mine owners, and Compass Minerals for supporting this project and, no less important, Jason Pepper for his invaluable IT support.

My greatest praise must go to my partner, Rachel Smith, who never baulked at carrying, standing neck-deep in water and modelling with the patience of a saint. She also helped immensely with the post-production activities in terms of literary and imaging expertise. I am an extremely lucky person.

My sincere thanks go to you all.

SOUTH WALES

The valleys of South Wales have a rich heritage synonymous with the mining of coal and iron. Throughout their working years, the countless sites evoked profound emotions; the pride and comradeship of workers and heartbreak of families, all too often caught up in mining tragedies. Today, the land lies quiet and nature has reasserted its presence.

Yet secreted beneath these hills are also found some of the longest and deepest cave systems in the British Isles. These are wonderous places to visit, with gently dipping limestone beds, dancing waterways and a profusion of beautiful decorations.

1	Ogof Gofan	11	Dinas Silica Mine
2	Llygad Llwchwr	12	Ogof Rhyd Sych
3	Dan yr Ogof	13	Ogof Agen Allwedd
4	Pwll Dwfn	14	Eglwys Faen
5	Tunnel Cave	15	Ogof Daren Cilau
6	Ogof Ffynnon Ddu 1	16	Ogof Craig a Ffynnon
7	Ogof Ffynnon Ddu 2	17	Ogof Capel
8	Pant Mawr Pot	18	Blaenavon Coal Mines
9	Little Neath River Cave	19	Ogof Draenen
10	Porth yr Ogof	20	Garth Iron Mines

Left The Hepste Resurgence in the headwaters of the River Neath.

HIDDEN REALMS

Ogof Gofan

SOUTH WALES

Ogof Gofan is unique in British caving. This cave lies well off the beaten track and a casual glance at its length and depth would leave anyone singularly unimpressed. But this is somewhere very special and provides a highly memorable day's activity.

Gofan lies halfway down spectacular limestone cliffs in the Pembrokeshire Coast National Park. It is also situated in an active military firing range! Permissions are necessary prior to visiting and it's a fascinating walk across the range to Saddle Head. Atlantic rollers batter the rocks far below and just everything about this place is breathtakingly invigorating. As ropes are set, climbers look on bemused with little or no inkling of the wonders lying out of sight beneath ground level.

Finding the abseil point is the first challenge as the entrance cannot be seen from above. The rocks are razor sharp and you really wouldn't want to drop anything. Once into the narrow, slippery cleft, a crawl leads to a wonderful oval window overlooking the sea. After this, a squeeze leads to a further low section and then ... you enter one of the finest speleological caverns in the UK. The chamber is adorned with flowstone of all sorts on every side, and against the far wall lies a small and beautiful clear green 'lake' which looks like a magical wishing well.

There is more beyond, and everyone who has been to Ogof Gofan will confirm that a trip into this cave is a glorious day out.

Left Main chamber.
Above Window over the sea.

HIDDEN REALMS

Llygad Llwchwr

Other than to a Welsh speaker the name of this cave is nearly impossible to pronounce. 'Lugad Lechweer' is reasonably close, but it's simpler and easier just to use the letters 'LL'. Translated, the name means the 'Eye of the Loughor'. This cave may not be long (1,200 metres), but it holds a special place in history as this was the first recorded site of sporting cave exploration in the Welsh underworld.

At LL a substantial stream flows from a fine oval entrance, a volume of water which gives every indication of a massive cave network stretching back into the mountain. While the water rises to the surface via a flooded shaft, the cavers' entry is via a small body-sized hole set about three metres above water level.

It's amazing to consider how, in the 1840s, the first explorers found their way through the intricate set of cracks and vertical fissures equipped as they were with tweed jackets, candles and oil lamps. They literally unrolled a ball of twine and the necessity for this soon becomes apparent to the visitor today. This place is complicated. Occasionally, the sound of running water leads one to a waterway where surprisingly the flow is seemingly from the wrong direction. Thomas Jenkins and his friends explored the entire length of the dry cave and revealed four river chambers, along one of which he took a collapsible boat, a coracle, to see where the river went. Audacious!

In recent years, divers have explored beyond River Chamber 4, including some sections of dry passage, but the many kilometres of undiscovered cave that assuredly exist currently elude all comers.

Above Deep water in River Chamber 2. **Right** A pillar formation near the entrance.

Dan yr Ogof

Dan yr Ogof is perhaps my favourite cave and, as Wales's only show cave, it is well known throughout the country and the UK. In 1912 the fabled Morgan brothers explored this notable site to the limits of existing equipment, turning back only when they met a daunting series of deep lakes and thunderous cascades. It wasn't until 1937 that cavers passed these to uncover a wealth of dry passageways beyond. Following their intuition and a good flow of fresh air, they passed numerous fine grottoes leading eventually to an intimidatingly tight and claustrophobic tunnel dubbed the Long Crawl. Here exploration halted until Easter 1966 when the squeezes were passed. The feat made headlines across the UK as the tunnels beyond were breathtakingly spectacular. Rarely will a caver experience such variety of terrain and speleological features.

Names such as the Grand Canyon, the Abyss, Bakerloo Straight and Flabbergasm Chasm give a clear, apt indication of places that excited those first explorers as they strode ecstatically along. Today, Cloud Chamber is regularly visited by cavers and enterprising television crews as this immense void is a wonderous sight by any standards. Stalactites and stalagmites are profuse, and pathways are carefully set out between lines of conservation tape. Beyond lies the Green Canal, where a thirty-to-forty-metre swim in chilly, eight-degree-Celsius water leads either to a return via a fine lower-level passage or a sporting continuation leading to the isolated and remote Far North.

Over seventeen kilometres of passages are currently known.

Right Cloud Chamber.

SOUTH WALES

HIDDEN REALMS

Pwll Dwfn

In a region where the caves are predominantly horizontal, requiring relatively little in the way of ropes or ladders, Pwll Dwfn is significantly different. It lies in a shallow, dry valley high on the Black Mountain and its claim to fame is that it is Wales's deepest natural pothole. The entrance is easily missed. It's an inconspicuous, body-sized hole giving little indication of what might lie beneath.

Pwll Dwfn translates as the 'Deep Hole' and, given its location, it is tempting to think that it might just drop into a subterranean branch of Dan yr Ogof which lies approximately one kilometre away. No other pothole in Wales compares, and a series of drops follow in quick succession meaning that the proficient, well-equipped caver will descend ninety-three metres rapidly. A set of five, predominantly dry, pitches provides a varied trip where the caver encounters beautiful shafts, an exciting pendulum and airy pitch-heads; an ideal place to practise the technical skills of SRT (single rope technique). With virtually no

horizontal passage between them, the descent leads straight down to water. Just when you think the cave will open to yield a horizontal streamway, it closes in at a pool and, a few metres distant, a sump. Whether this site will ever be passed remains to be seen.

The drops leading to the bottom present a regular attraction for British cavers; Pwll Dwfn is arguably the finest vertical site in Wales, indeed in the southern part of Great Britain.

Above left The entrance. **Above right** The second pitch. **Right** The third pitch.

HIDDEN REALMS

Tunnel Cave

Tunnel Cave is noteworthy in that it is a system of two parts. This cave was first explored from the bottom of the mountain, from what is today the tourist attraction of Cathedral Cave – a site renamed when it was incorporated into the Dan yr Ogof show cave network in 1971. The second, upper entrance is situated high above on the western flank of the Swansea Valley.

Tunnel Top – what a breathtaking place to start and end a day's caving. You lift a heavy metal lid, set flush at ground level, and peer down a vertical shaft. This twelve-metre drop was mined through solid rock in 1963 – a bold operation by local explorers to intercept a high point in the system at a time when access was restricted at the original entry point. A second eight-metre drop leads to the head of a spacious, steeply sloping ramp. This is an impressive place. The cave drops rapidly until you find yourself in a tall and winding fissure of reduced proportions which, with frequent traversing at roof level, leads to the eventual termination, the artificial barrier set up at the end of Cathedral Cave. At the present time through trips are not permitted, but a visit to the public section is thoroughly recommended: the renaming to Cathedral Cave is apt, and with its grand walkway and high waterfall this is certainly one of the more spectacular show caves in the British Isles.

Above left Bottom of the first pitch. **Above right** Waterfall in Cathedral Cave.
Right Rope traverse near the upper entrance.

HIDDEN REALMS

Ogof Ffynnon Ddu 1

Located at the head of the Swansea Valley is the complex system of Ogof Ffynnon Ddu, or OFD. The cave extends to over sixty kilometres in length with a depth range in excess of 274 metres. It is one of the most popular caves in the UK with trips to suit all levels of ability and accessible in all weather conditions.

For convenience, cavers think of the cave system in three sections: OFD 1, 2 and 3, all of which were originally separated by sumps. OFD 1 lies close to the resurgence at the valley bottom and was first entered in 1946.

So, what makes OFD 1 so great? Very simply, the magnificent streamway which is generally held to be the finest stream passage in the British Isles. The waterway, which drains to the resurgence, is easy walking and follows a tall, canyon-like passage, well sculpted and clean washed. With a gentle gradient, the water is rarely more than thigh-deep, unless you stumble into one of the notorious, deeper swirl holes or 'pots'. Anything that was potentially dangerous has been tamed by the introduction of scaffold bars to cross the pots, and in the event of rising water a high-level flood escape route has been established involving some airy traversing along well-maintained wire cables in the roof of the canyon. Always popular is the 'round trip' which takes you along the streamway, up Lowe's Climb and out via the large and impressive upper series.

Above Pi Chamber. **Right** Crossing a flooded pot in low water conditions.

HIDDEN REALMS

Ogof Ffynnon Ddu 2

Ogof Ffynnon Ddu 2 is located higher up the mountain than OFD 1 and is accessed via either the Cwm Dwr Quarry Cave or the Top Entrance, a kilometre or so even further up the hill.

OFD 2 was not discovered until a diving breakthrough at the upstream end of OFD 1 in 1966. Within a short time, the cave system became one of the longest and deepest in the British Isles. Trips into this section can be incredibly varied with popular circuits taking in many of the stunning formations and spectacular grottoes. Surprisingly, for a cave of such length and depth, there are relatively few pitches, and the vast majority of the system can be traversed without having to resort to vertical equipment. Trips from OFD 1 to OFD 2, or from Cwm Dwr to OFD 1 or the Top Entrance, are classic undertakings, but it must be said that route finding is challenging.

Today, there is yet another section of cave, namely OFD 3, the furthermost upstream sector of the complex which is normally accessed via Top Entrance. This is a more challenging trip than the others and does require vertical equipment. The traverses leading into OFD 3 are particularly airy and exposed.

OFD presents a cave of the highest calibre. There is little doubt that you will leave wanting to see more. It is without question a fabulous system and deservedly designated a Site of Special Scientific Interest and a National Nature Reserve.

Above The Columns. **Right** Skyhook Passage.

Pant Mawr Pot

To reach Pant Mawr Pot requires a four-kilometre trek. It is located in a remote area of moorland roughly midway between the Neath and Swansea valleys. This cave is accessed via a sixteen-metre-deep shaft, first bottomed in 1936, and in 1953 it was further extended to its current length of 1,100 metres by members of the South Wales Caving Club.

At the foot of the daylight shaft, one alights in a large sloping cavern with the sound of flowing water. The upstream passage is short, but downstream a spacious tunnel can be followed through mixed terrain, occasionally clambering through massive boulders, to an eventual narrowing and sump. The water from the cave reappears nearly four kilometres away on the western bank of the Nedd Valley. There is little doubt that a major continuation of Pant Mawr awaits explorers in the fullness of time.

Along the way there are some large and very fine stalactite draperies hanging over the stream passage. Smaller, but no less beautiful, are the delicate straws and exquisite arrays of helictite formations set discretely off to the side in places such as the Chapel. The cave presents a really good sporting day out and you will emerge surprisingly clean. Remember to allow ample time for your trip; inevitably it will be a long day. It is certainly a fine collector's piece.

Top right Helictite formations in the Chapel. **Right** The entrance shaft. **Opposite** Sabre Junction.

HIDDEN REALMS

Little Neath River Cave

SOUTH WALES

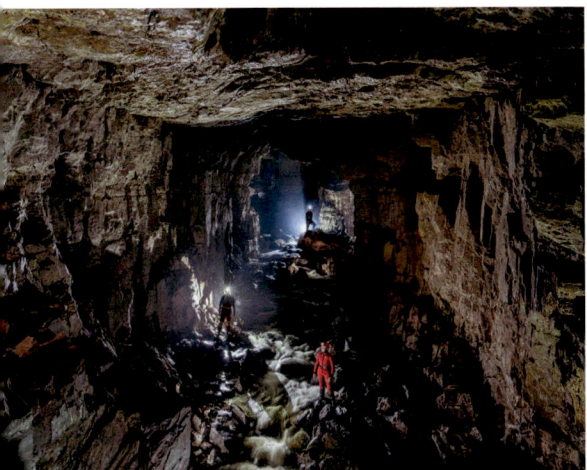

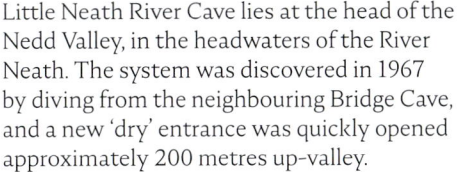

Left A grotto in the Oxbow.
Below The main passage of Bridge Cave.
Bottom The entrance in flood.
Right Formations next to the streamway.

Little Neath River Cave lies at the head of the Nedd Valley, in the headwaters of the River Neath. The system was discovered in 1967 by diving from the neighbouring Bridge Cave, and a new 'dry' entrance was quickly opened approximately 200 metres up-valley.

The location of the small entrance will raise eyebrows for anyone who has never been there before. It is sited directly in the bed of the small upland river. Previous visitors have constructed a low dam of cobbles, which has highly dubious qualities at holding back the ingress of water to the cave.

The first fifty metres of this eight-kilometre-long cave are undoubtedly the most memorable. Lying flat out in the streamway, it doesn't take much imagination to see that the place could fill with water in a matter of minutes. Going into the cave, with the flow carrying you forwards, is one thing. Coming out with the strong flow in your face is another matter. Under summer conditions the water feels like a lukewarm bath; in winter the water levels are higher and the temperature bracing.

Once inside the cave the passages are generally substantial excepting a flat-out section where one drifts along in the waterway – the Canal. Turning left at the end of this presents a fine circuit, with a lovely grotto en route back towards the entrance. Continuing down the river leads via an increasingly impressive tunnel to the terminal sump. This cave proved such a wonderful day's adventure that their Royal Highnesses Princes William and Harry visited in two consecutive years.

HIDDEN REALMS

Porth yr Ogof

Porth yr Ogof lies at the very heart of the Brecon Beacons National Park. This is such a wonderful place to discover the underground world and it has slowly evolved to become the premier induction site for novices in the UK. It is for good reason that some 30,000 people a year come here to experience caving for the first time: this site has it all. The passageways are, in general, big and easy to move through; you can deviate to enjoy sporting challenges of various kinds; there are formations and water in abundance.

The grand Main Entrance captivated travellers from the earliest of times. It has inspired notable artwork and generated fascinating folklore. There is much to see, from the quietly swirling waters of the fabled White Horse Pool to the vast expanse of unsupported roof in the Great Bedding Cave. And there is mystery, especially regarding the flooded 'diver only' sections of cave, where exploration still beckons tantalisingly.

The power of water is evident throughout, and seeing huge tree trunks wedged across a passage and smaller items of driftwood lodged in the highest cracks in the roof leaves one with a healthy respect for the elemental power of the Mellte river. This is a very special place with opportunities for everyone; plan your visit for a dry sunny day and you will thoroughly enjoy this superb cave.

Above Hywel's Grotto. **Right** White Horse Pool.

SOUTH WALES

HIDDEN REALMS

Dinas Silica Mine

Above Winding drum in the upper levels.   **Opposite** Diver in a flooded level.

The Dinas Silica Mine nestles at the head of the Sychryd Gorge. The scenery in this headwater of the River Neath is spectacular, with massive overhanging cliffs, cascades and waterfalls. The mine sets it off and I thoroughly commend a visit to anyone in the area.

Mine operations ceased in 1962, before which this was an important place for the extraction of silica. The high-quality stone was crushed and used in the production of fire bricks to line the walls of the blast furnaces located further down the valley.

Walk into the tunnels and you step back into a historical environment where, in the nineteenth and twentieth centuries, miners used the typical pillar and stall technique of working. Very simply, the miners took all the rock that they could safely remove, leaving a network of pillars to support the roof.

Scores of men worked here at the height of operations, extending the mine about 400 metres into the hill along a series of parallel, interconnected levels. Artefacts are limited, but search the upper part and an array of interesting features will be found. There are a lot of tunnels in the dry part of the network and just as much again in that part of the complex which today lies below water. Interestingly, the flooded tunnels are one of the most important cave diving training sites in the British Isles and divers will regularly be seen operating here.

HIDDEN REALMS

Ogof Rhyd Sych

Ogof Rhyd Sych is an adventure. The cave begins in a wonderful mossy gorge, and walking into the clean-washed passage you would imagine that things would all be plain sailing. But this is a cave which changes rapidly, first to an intimidating-looking duck, and then into a seemingly complex navigation through low, gently inclined bedding planes.

You will want to have checked the weather forecast before visiting this site because once you are lying flat out, wondering if you are wriggling through at the easiest, most spacious bit, you really don't want to be thinking of rain above ground. This place would be an absolute horror show were you to be in the cave with the water rising.

From one bedding plane to the next, then into an equally constricted vertical fissure, then back to bedding plane ... the entertainment requires a positive mental attitude and a good pair of knee pads in the oft-gnarly sections.

But when you suddenly break free into a walking-size passage and see the formations: wow! The place is superb – what a transformation. There follows a huge tunnel to the eventual boulder choke, and arriving there you really feel you have achieved something. This 1,100-metre system may not be the longest cave in the Glais Valley, but it certainly provides one of the most memorable caving experiences in Wales.

The site was first investigated in 1950 but the major extensions fell to members of the Cwmbran Caving Club in 1967.

Right The first grotto.

HIDDEN REALMS

Ogof Agen Allwedd

SOUTH WALES

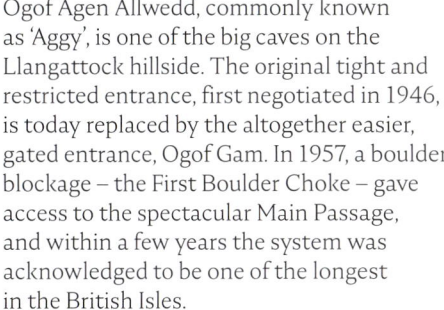

Ogof Agen Allwedd, commonly known as 'Aggy', is one of the big caves on the Llangattock hillside. The original tight and restricted entrance, first negotiated in 1946, is today replaced by the altogether easier, gated entrance, Ogof Gam. In 1957, a boulder blockage – the First Boulder Choke – gave access to the spectacular Main Passage, and within a few years the system was acknowledged to be one of the longest in the British Isles.

The impressive Main Passage presents a leisurely romp for the best part of a kilometre. Its spacious dimensions provide an oft-quoted indicator of size when cavers draw comparisons with sections of tunnel elsewhere in the UK. The route undulates gently along a hard-packed, dry clay floor which is crazed with fissures and interspersed by areas of rocks and breakdown. Occasionally, areas to the side are taped off to preserve delicate selenite crystals which seem to grow from the floor. Here, too, any number of bats (generally lesser horseshoe) will be found hibernating over the winter months.

To the side of this wonderful tunnel a selection of challenging passages lead deeper into the mountain. For the dedicated caver there lies the possibility of a couple of impressively long circuits. These are trips in excess of six hours duration, even assuming that the leader is confident with their navigation. The tunnels of the westernmost Summertime Series are spectacularly large, and lovely grottoes will be found along Turkey Streamway.

In all, Aggy presents over thirty-four kilometres of passages with the prospect of significant future additions. In the further reaches, the cave lies perhaps fifteen metres from neighbouring Ogof Daren Cilau, to which it must surely be connected in the fullness of time.

Left Southern Stream Junction.   **Above** Selenite crystals in the Main Passage.   **Above right** The Courtesan.

HIDDEN REALMS

Eglwys Faen

Eglwys Faen occupies a highly conspicuous position in the Craig y Cilau National Nature Reserve on the Llangattock hillside. Historically, this cave was one of the first to be entered in Wales, and while the walls have succumbed to much graffiti, those very inscriptions give some fascinating insights. The quarrymen who revealed a number of entrances along this dramatic escarpment clearly sought shelter inside the spacious Main Chamber when the weather was dire, but it is also interesting to ponder the fact that the very name of this cave, which translated means the 'Stone Church', hints at something even more intriguing. Research suggests that this was the site of religious services during the Civil War of the seventeenth century. Today, this is a popular venue with outdoor providers and an ideal site for novice activities.

The large Main Chamber, accessed directly from the surface, is particularly grand and a clear indicator of an extensive, currently unknown, system somewhere further into the hillside. Cavers achieved a significant advance in 1956 when a protracted dig revealed the Inner Chamber and the Warren a few metres above, but despite intensive excavational projects over the years the site has never divulged its innermost secrets. Today, the passages extend to 1,300 metres. The enticing draughts blowing through these tunnels tell the story: there is a lot of cave waiting to be explored here with possible connections with both Ogof Agen Allwedd and Ogof Daren Cilau.

Above The Warren, upper series.   **Right** Main Chamber waterfall.

HIDDEN REALMS

Ogof Daren Cilau

Ogof Daren Cilau (pronounced 'Daren kill-eye') lies in an abandoned quarry on the Llangattock hillside. Everything about the place is challenging, and the entrance series is downright intimidating. It starts as a wet, flat-out wriggle and that's how it continues for the next 517 metres; there are also a couple of notable squeezes on the way and a chilly draught to ensure that you don't hang around.

When you break free of the entrance passage, and after you pass the 1984 breakthrough boulder choke, you embark upon a joyous adventure. Striding along, on and on, you begin to grasp the size of this cave. In one direction you traverse Epocalypse Way, passing pristine formations – Urchin Oxbow and the famous group of helictites, the Antlers.

In the other direction is a significantly longer trip towards the terminal downstream sump. This will take you to the monumentally large tunnel, the Time Machine, and eventually to the 'camp' at Hard Rock Cafe, so it's not a venture for the unfit or faint-hearted. From here, even longer undertakings are possible towards the Restaurant at the End of the Universe and those far-flung sections of cave close to the neighbouring system Ogof Agen Allwedd; these are multi-day affairs. What an incredible world-class traverse will be on offer when these two caves are eventually connected.

Above left A Pin Cushion formation in Urchin Oxbow. **Above right** Psychotronic Strangeways. **Right** The Vice squeeze in the entrance passage.

HIDDEN REALMS

Ogof Craig a Ffynnon

Ogof Craig a Ffynnon lies in an old quarry on the north side of the Clydach Gorge close to the village of Blackrock. First entered in 1976, the system currently extends to eight kilometres but inevitably a connection will be achieved with other major caves beneath the Llangattock hillside.

The streamway a short distance inside does flood, but that aside the cave is accessible at most times of the year. There are plenty of formations throughout and it is to the credit of cavers that the wonders of this site remain so well conserved.

Passing along the main route you cannot fail to be impressed by the sheer effort that has gone into exploration here. The First Boulder Choke has collapsed several times and, given the small stream tumbling through the rocks, this is perhaps understandable. A really protracted effort finally sees you pass through the Second Boulder Choke and, while twisting, small and muddy, this place is good and safe. Thereafter the cave opens out into a huge fossil tunnel and, while you may have to wallow unceremoniously through the 'land of gloop', you will soon be awestruck by a succession of fine scenes that appear one after another. The most spectacular is undoubtedly the Hall of the Mountain King, but pass beyond the Third and Fourth Boulder Chokes and other exquisite sights will be encountered. Your kit may require a good wash when you finally emerge, but you will have enjoyed every moment.

Left Hall of the Mountain King. **Above** Display of straws.

HIDDEN REALMS

Ogof Capel

Ogof Capel is a lovely little resurgence cave located in a crag directly below the Heads of the Valleys Road at Blackrock. The original hobbit-sized entrance leads quickly to a small clear-water sump; the early exploration was conducted by cave divers who passed two short sumps to reach a massive area of boulder collapse, directly beneath the busy trunk road!

Today, a dry route, via an excavation in neighbouring Ogof Gelynnen, has been established which bypasses the flooded sections. Cavers still have to negotiate a perilous boulder collapse where, despite some shoring, care is required. Once through, you emerge to a comfortably sized passage heading straight back into the hillside. The place is a little gem and there are formations at every turn: stalactites, stalagmites, curtains and helictites in glorious profusion. It takes time to negotiate this cave as you have to move slowly to avoid damage; there are many short deviations to protect incredibly well-decorated sections.

There's a momentary change of character where a low canal-cum-duck takes you through a normal bit of cave passage, but thereafter it's back to the same ... wow, for an 800-metre-long cave this place is outstanding.

Left Broken Stalactite Grotto. **Below** Straw cluster.

HIDDEN REALMS

Blaenavon Coal Mines

From the very beginning of the Industrial Revolution coal was mined in the Blaenavon area. Here, at the north-eastern rim of the South Wales Coalfield, deposits outcrop at the surface and seams dip away ever deeper to the south. Mining commenced here in the late eighteenth century and as the trade became increasingly important colliers were forced to tunnel further and deeper to obtain their rewards.

Today, coal is no longer produced near Blaenavon, but a number of abandoned mines can be found in the valley. Much of the machinery and infrastructure was left behind, but these sites have not been inspected or maintained since their closure, making them highly dangerous due to instability and the risk of bad air. These are extremely perilous places.

The exception to this is Big Pit, which is maintained as a show mine by Amgueddfa Cymru – Museum Wales. Mining started at Big Pit in about 1880 and during its working life material was extracted from nine separate seams. At the height of operations, 1,300 men were employed and produced over 250,000 tons of steam coal per year, primarily for use on the railways. Production ceased here in 1980 and it subsequently reopened as a visitor centre in 1983. Today, over 100,000 people a year pass through the complex. Equipped with helmet, belt, lamp and battery pack, visitors descend in the miners' cage to the bottom of a ninety-metre shaft. Here they are given a wonderful insight into a fascinating bygone age and a good idea of the once daily way of life for tens of thousands of workers beneath the hills and valleys of South Wales.

Above Big Pit mine complex. **Right** Main tunnel in an abandoned mine.

HIDDEN REALMS

Ogof Draenen

Until 1994 Ogof Draenen was no more than a draughting hole in a post-industrial mountainside. Set between Gilwern Hill and the Blorenge, just north of Blaenavon, who could have imagined what wonders would be uncovered in this locality. After a lengthy excavation, scaffolding and shoring as they went, members of Morgannwg Caving Club dropped through into a walking-sized rift ...

Within a few metres there was a gaping hole into blackness. Big Bang Pitch was laddered a few days later and large passages were found leading off in virtually all directions. Ogof Draenen today is the longest continuous, dry complex in the British Isles with approximately seventy kilometres of passages; visitors are advised to take a survey and prepare well!

Routes were quickly surveyed under Gilwern Hill and in the other direction towards the water's exit from the mountain near Pontypool. With passages plotted, the cave quickly reached the subterranean edge of Blaenavon and draughting tunnels gave indication of much more to follow. Other entrances were opened in later years. The cave presents spectacular, complex trips and circuits to suit all comers.

One classic journey is to the Dollimore Series. This area is remote but it's an incredible experience to see the wonders of Circus Maximus, the Geryon and others; these formations are quite, quite breathtaking.

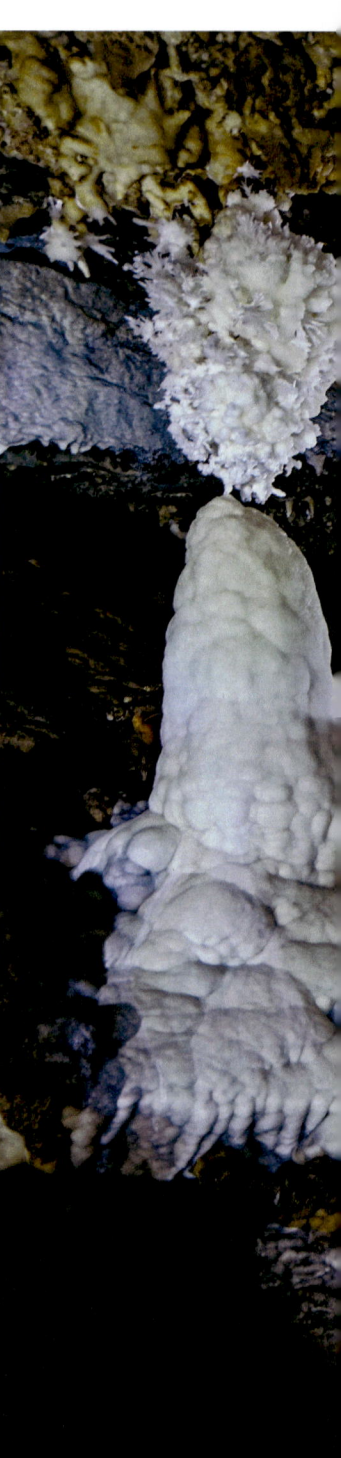

Above The Geryon. **Right** Formations near Circus Maximus.

HIDDEN REALMS

Garth Iron Mines

Garth Iron Mines are located in Garth Wood, immediately north of Taffs Well Quarry near Cardiff. While not a lengthy complex, the abandoned workings extend to a depth in excess of 100 metres.

Iron mining in the area can be traced as far back as Roman times, while historical records show underground activity here dating to 1560. By the mid-1800s some substantial caverns had been excavated – so much so that Garth was known as a tourist attraction in Edwardian and Victorian times! The mine was finally abandoned in 1937 and the lower levels of the site were left to flood. Of the resulting four lakes, one is known to have a depth of at least fifty-five metres. During World War II there was a period of time when parts of the mine were used as an ammunition store.

Throughout the mine there are vertical sections and steep slopes between levels. In one place, sunlight permeates to the lowest point, illuminating the ochre-coloured walls that surround and dip into the depths of a blue-green lake. No one could fail to be impressed by the colourful nature and the sheer size of the chambers in Garth. There is little doubt that the caverns existing here are some of the largest in the UK. This is, in a very real sense, an awesome place.

Left Footpath along a ledge.   **Above** Impressive excavations in upper levels.

MID AND NORTH WALES

The mountains and hills of Mid and North Wales have long been quarried and mined for slate and mineral ores. Look in the right places and you will quickly discover evidence of this activity: spoil heaps and the ruins of old buildings, rusty machinery, and long-abandoned grassy inclines.

Beneath the surface lie great dark voids, often with fathomless blue lakes – clear indicators of deep flooded levels below. There are trams and rails, colourful mineral deposits (copper, lead and zinc), and sometimes more personal artefacts left behind by the people who worked these sites. This is a fascinating region to explore.

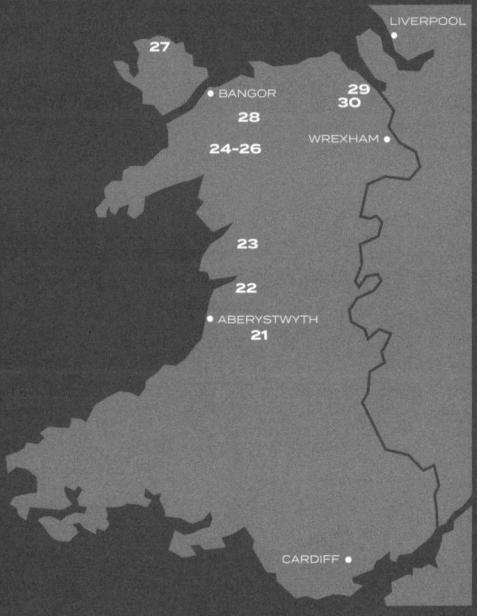

21	Level Fawr Lead Mine
22	Ystrad Einion Copper Mine
23	Aberllefenni Slate Mine
24	Rhiw-bach Slate Mine
25	Cwmorthin Slate Mine
26	Croesor–Rhosydd Slate Mine
27	Parys Mountain Copper Mines
28	Parc Lead Mine
29	Milwr Tunnel
30	Ogof Hesp Alyn

Left Surface ruins at the Croesor–Rhosydd Slate Mine.

HIDDEN REALMS

Level Fawr Lead Mine

Level Fawr Lead Mine, which is also known as Bonsall's Level, lies in the Cwmystwyth Valley lead mining complex between Aberystwyth and Rhayader in Mid Wales. This area feels desolate and remote. Few people live in the locality today, a far cry from the early nineteenth century when this was one of the foremost lead mining areas in the world.

While evidence has been found of lead working in the Bronze Age, it is believed that the Romans were the first to work the site intensively. The mine reached its zenith in the 1760s under the direction of Thomas Bonsall, who came from Derbyshire. It remained an important site until there was a massive decline in mining in the twentieth century; by World War II, the fate of the industry was sealed.

Level Fawr itself extends for more than three and a half kilometres and you will quickly find that this is a fascinating place. The entrance area has been stabilised by mine enthusiasts in recent times, but beyond this there are many potential hazards. There is plenty to see: a sizeable old tram set upon rusty rails, ore chutes, decaying wooden ladders disappearing into the heights; indeed, all manner of artefacts. Ascend with caution to higher levels and you will find enormous chambers painstakingly excavated by laborious effort, loose rubble slopes and plenty of drops. It's an amazing place.

Today, a few ruins dot the abundant spoil heaps and it's not immediately apparent where the entrance to Level Fawr lies. Whatever opening you chance upon first, take care. All are potentially unstable.

Above left Skipway incline. **Above right** Roof supports. **Right** Ore chute.

HIDDEN REALMS

Ystrad Einion Copper Mine

Ystrad Einion Copper Mine lies at the head of Artists Valley above the village of Furnace, north of Aberystwyth. Other than the information board and a few ruined structures, little remains on the surface to indicate that mining once took place here. A diligent search amid the forestry will reveal that entrances to the mine exist on more than one level, but the main entrance is marked with a substantial metal gate and has a very small stream flowing out.

Stepping carefully should ensure that water does not get into your wellies, and an easy walking tunnel leads in for a relatively short distance to an outstanding sight. Here, the miners constructed a massive underground waterwheel, one of only two to be found anywhere in the UK. Sadly, this is now falling into decay, but suffice it to say that it makes a spectacular sight. Close by lies a deep flooded shaft, while exploring another tunnel will reveal a five-metre climb leading to yet another interesting relic of a long bygone age: a large metal hauling container for the ore – a kibble. One presumes that the ore body at Ystrad Einion was of limited extent, and little evidence of copper, or indeed any other mineral, is readily apparent today. But seeing a waterwheel and a substantial kibble *in situ* is particularly special. Such artefacts must be treated with all due respect.

Top right The waterwheel. **Right** Entrance tunnel. **Opposite** Plank crossing deep water.

HIDDEN REALMS

Aberllefenni Slate Mine

MID AND NORTH WALES

Opposite Large daylight cavern.
Left Ladder between levels.
Below Dinghy in flooded cavern.

Aberllefenni Slate Mine is a spectacular site in every respect. The mine was evidently a significant producer of slate during the Industrial Revolution but finally stopped working around 2000 when the miners closed the doors at valley bottom for good. Today, that entrance has been sealed, but find a way in and you will be amazed at what has been left behind.

The primary level that leads right through the mountain is nothing special, but along that route huge caverns will be revealed. Some admit daylight from high above while others present yawning, perilous, gaping shafts off to the side. There are fathomless blue lakes of clear water and artefacts such as a lovely 'rock shoveller' and a set of trams. Directly adjacent, a massive metal crane bestrides a deep, flooded chamber. These features clearly date from the modern age and lie parked up as though awaiting the morning shift to arrive.

Aberllefenni Slate Mine lies a short distance from Corris, near Machynlleth. It is worthy of a visit at any time, but the place seems all the more worthwhile when the weather is dire and you can't appreciate the glorious scenery on the surface. It's a real treasure in the subterranean world of Mid Wales.

HIDDEN REALMS

Rhiw-bach Slate Mine

Rhiw-bach Slate Mine is an absolute delight and probably one of the safest mines to visit in the UK. As such, it is equipped and regularly in use by adventure activity providers. The mine is located at the head of the Machno Valley above the old quarrying village of Penmachno.

It's a steep haul of about one and a half kilometres to the top of the mountain, but the scenery is wonderful. Here the gated entrance to the mine within the Rhiw-bach Quarry is impressively substantial and inside you step back into history. The multiplicity of small-diameter, drilled holes gives a clear indication as to how long and hard it was to win the slate in the earliest years of extraction. An old rock cart is remarkably well preserved here. Very quickly you join a small trickle of a stream which would have collected and then conducted water down and away from the working area. As you follow this, the route is intercepted by levels and caverns off to the side.

The way through to the bottom entrance is 'dry' but passes fascinating areas of deep water: either a fabulous clear blue or, at the lower end of the complex, somewhat murky due to the ferrying of groups across a lake in a dinghy. When you see a little speck of light down the tunnel you think you are nearly out, but amazingly that bottom entrance is more than a little way off – it's a solid ten-minute walk away. This is a fine and highly memorable place.

Above Blue lake cavern. **Right** Dinghy in a partially flooded cavern.

MID AND NORTH WALES

HIDDEN REALMS

Cwmorthin Slate Mine

Quarrying began at Cwmorthin Slate Mine in 1810 and in the nineteenth century it was one of the most important sites in the world. The mine ceased operations in 1997 and, along with five other slate mines in Gwynedd, was designated a UNESCO World Heritage Site in 2021.

The entrance is an easy walk from the car parking area, located close to Tanygrisiau in Snowdonia. Given the many kilometres to which the complex extends underground, there are a variety of trips to suit all levels of ability. Descending into the depths it is interesting to reflect upon the scale of past working operations. Some of the galleries are huge and care is always required, especially moving through cold, potentially unstable areas where it is wet and slippery. There are many industrial artefacts scattered throughout the mine.

Today, Cwmorthin is a very popular site with members of the public and outdoor groups. Adventurous activities conducted here are coordinated by a professional company with qualified leaders and appropriate safety equipment; an exciting and different way to see North Wales. To quote from Go Below Underground Adventures Ltd: 'Cross ancient bridges, traverse catwalks and miners' stairways to reach a point that is 1,375 vertical feet underground – the very deepest place in the UK accessible to the public.'

This is a pretty unique place and certainly commercially this can claim to be the 'ultimate underground adventure' in the UK.

Above left Abseiling inside the mine. **Above right** An airy high-level catwalk. **Right** Water crossing and trucks.

HIDDEN REALMS

Croesor–Rhosydd Slate Mine

MID AND NORTH WALES

The Croesor–Rhosydd mine network is outstanding. The labyrinth of tunnels lies high on the mountain between Tanygrisiau to the east and the small village of Croesor to the west. In essence, there are two separate mine entrances one kilometre apart and it's a steady, stiff ascent to reach the place from either side.

Since 2000, this complex site has seen considerable attention, both by keen local activists and by a growing number of visitors enthralled and attracted by the nature of the place. A trip here is a particularly challenging undertaking. The traverse from one mine to the other has it all. It's physically and mentally demanding; it requires technical knowhow; you need to be thoroughly prepared for what could prove to be a long, cold day and you need to be both careful and constantly alert. This is a mine where you need to navigate carefully, where the terrain underfoot is often unstable and significant rockfalls are a regular occurrence. Safety must be the primary consideration at all times.

Despite these words of warning, this is a fantastic trip. Starting from Croesor, the journey through the hillside may take some four hours and there is so much to see and do along the way. There are abseils, short scrambles and climbs, and areas of deep water, together with permanent and temporary aids such as Tyrolean crossings, bridges and a canoe. With due care, good planning and suitable equipment you will emerge at the far side elated by your day's adventure.

Left Tyrolean crossing over a deep lake.   **Above** Enormous chamber in the mine.

HIDDEN REALMS

Parys Mountain Copper Mines

Parys Mountain Copper Mines, on Anglesey, deserve a visit if only to view the huge and colourful open-cast pit from which much of the valuable ore was extracted.

Copper deposits were mined here from as early as 1900 BC. Many interesting archaeological discoveries have been made and perhaps the most fascinating are that of a Bronze Age bell pit – initially uncovered by Victorian miners – and a quantity of rounded stone pounders similar to those found in the copper mines of the Great Orme complex near Llandudno.

In the 1780s the Parys mines were to yield the highest tonnage of ore in the world, some 3,000 tons per annum, and the workforce was an astounding 1,500 men, women and children. Copper was eventually worked to a vertical range of 270 metres, the greatest depth achievable using the technologies of the time. Ore was excavated from shafts and levels; later on, much of the site was worked by open-cast methods. Large-scale mining operations continued up to 1883 and all mining seems to have ceased in the early years of the twentieth century.

The mine contains some beautiful bright-blue formations, coloured by copper, and an excellently preserved mid-nineteenth-century wheelbarrow of Cornish design. Most rewardingly, visitors can choose to exit via an aqueous connection with the neighbouring Mona Mine. Thanks to the 'dewatering' of the lower sections in 2003, around eight kilometres of tunnels are accessible today; a truly fascinating and historic place.

Left The crater on Parys Mountain.
Above Wheelbarrow and besom broom.
Right Deep red lake.

HIDDEN REALMS

Parc Lead Mine

Parc Lead Mine, near Llanrwst, was the last working mine in the Gwydyr Forest. With its connections to the older Llanrwst and Cyffty mines, this is an important site for its history and industrial archaeology. Workings commenced in the early seventeenth century, but the heyday of mining was the period from 1860 until 1900. Both lead and zinc ores were extracted and processed, but often the financial returns were poor. After World War II, prospects improved, and with more modern equipment, a diesel locomotive and a better separation plant, the enterprise ran at a profit. The mine closed in 1963.

In 1968 a small area on Level 2 was used for a short-lived scientific study – the Bidston Experiment. This tried to measure the deformation of the earth's crust induced by tidal flows on the coast. Sensitive pendulums and ancillary equipment were installed, but the results were inconclusive.

Entry to the mine today is via the Level 2 portal, which leads to an easy walking tunnel, albeit with a number of perilous forty-metre-deep holes dropping to the underlying level below. About 750 metres from the entrance a pumping shaft is located, and this still has a large wooden pump rod *in situ*. The timbers in the upper workings are in poor condition and significant collapse has occurred over the years. Some ore chutes have suffered from decay, but many good surviving examples can be seen, along with various other fixtures and artefacts.

Top right Level 2 ladderway.
Right High stope and timbers on Level 2.
Opposite Rails and ore chutes on Level 2.

HIDDEN REALMS

Milwr Tunnel

Few places in the underground world are as interesting as the Milwr Tunnel. This site in North East Wales is famous for its sixteen-kilometre-long drainage conduit which dewaters the abandoned lead mines between Holywell and Mold. In 1897, driving of the tunnel began from sea level at Bagillt on the Dee Estuary. It was progressively extended inland up until 1957 when it reached its terminus near Loggerheads. As a direct consequence, the mines could be worked to substantially deeper levels than would otherwise have been possible. The workings closed in 1987 and today the Milwr Tunnel continues to disgorge a flow of twenty-three million gallons of water per day, rising to thirty-six million in wet weather.

One can descend around 150 metres to the level of the waterway at several points via a number of old mines. Amazing sights will be found, including at one spot a substantial old train with heavily rusted carriages designed to carry men through the mine. Further along the same side branch is a huge natural chamber – Powell's Lode – occupied by an enormous deep, clear-water sump; this is a phenomenal site in itself.

Another tunnel leads to gigantic underground limestone workings, below Hendre. The scale of the operation is breathtaking, and scattered around in this area is all manner of abandoned quarrying machinery including mechanical rock shovellers and trams. In total, this complex of mines, quarries and drainage tunnel extends to around 100 kilometres. This extremely impressive site is also one of the longest subterranean networks in the UK.

Right Powell's Lode Cavern.

MID AND NORTH WALES

HIDDEN REALMS

Ogof Hesp Alyn

Ogof Hesp Alyn (Cave of the dry Alyn) near the village of Loggerheads was first entered in 1973 by members of the North Wales Caving Club. This is a fascinating cave and quite unique, being entirely phreatic in nature. Prior to the nineteenth century, it was completely flooded and inaccessible to cavers. Originally the water from this system drained to a spring east of Cilcain, but when mining activities intercepted the deep, flooded passages the course of the water flow changed, and Ogof Hesp Alyn was drained. Today, the complex is around 1,400 metres long, with an overall depth of ninety-eight metres.

Given the entrance location, directly adjacent to the intermittent surface river, dry stable weather is essential for a trip here. This cave is full of challenges: vertical descents, climbs, a traverse and a duck, and all these are interposed with sections of exceptionally muddy passage. A good sense of humour will certainly assist your day!

Interestingly, you don't encounter any flowing water at all in the cave. Sumps block the route at the furthermost dry point and about 400 metres of muddy extensions are known beyond. So, where is the missing streamway that eventually cascades impressively into Llyn-y-Pandy Mine we ask? Undoubtedly, it's there somewhere along with a lot more passage, all simply waiting to be found ...

Above left Muddy phreatic tube.   **Above right** The duck.   **Right** Ladder climb in mud.

MENDIPS AND SOUTHERN ENGLAND

The limestone hills of the Mendips are famous for tourist sites such as Cheddar Gorge and Wookey Hole, but there are many more caves here, often richly decorated with fabulous formations. It is easy to gain depth without really noticing – small waterfalls, little climbs and steeply sloping passages. On your return you realise how far down you have been!

Travel onwards from the Mendips and Southern England has much more to reveal: stone mines near Bath, chalk mines near Reading, caves around Buckfastleigh and the extensive, colourful workings of Devon and Cornwall where surface buildings stand etched on the skyline – beautiful and iconic structures. There is so much history here ...

#	Site	#	Site
31	St Cuthbert's Swallet	37	Reservoir Hole
32	Swildon's Hole	38	Shatter Cave
33	Wookey Hole	39	Withyhill Cave
34	Upper Flood Swallet	40	Stoke Lane Slocker
35	GB Cave	41	Pen Park Hole
36	Charterhouse Cave	42	Bath Stone Mines
		43	Emmer Green Chalk Mine
		44	Pridhamsleigh Cavern
		45	Reed's Cave
		46	Tywarnhayle Copper Mine
		47	Rosevale Tin Mine

Left Cheddar Gorge.

St Cuthbert's Swallet

St Cuthbert's Swallet is a classic UK cave system, approximately seven kilometres in extent with a vertical range of 145 metres. Lying on the eastern edge of Priddy village, you don't have to walk far to reach the entrance. Your leader will be equipped with both a key, for the manhole cover gate, and also a hefty spanner which will be used to close a sluice gate to enable you to enter the small stream sink reasonably dry.

The concrete shaft leads quickly to an eight-metre-deep entrance rift. You peer down the narrow fissure and as the sluice closes the stream rapidly diminishes to a mere trickle. The short pitch is not so much tight as awkward, but that overcome, there follows a continued rapid descent on fixed metal ladders. You soon realise that the place is a vertical warren – intricate, fascinating and very soon superbly decorated. St Cuthbert's is known especially for its magnificent curtains and calcite flows which are found at many locations throughout the cave.

Since the cave was entered in 1953 it has witnessed tremendous efforts to try and reach the elusive passages that assuredly run three kilometres beneath the Mendips to the resurgence at Wookey Hole.

Any number of interesting circuits are possible in this complex system, taking in countless wonderful sights along the way.

Above Curtain Chamber.   **Right** Flowstone cascade.

HIDDEN REALMS

Swildon's Hole

First entered in 1901, this place is always an absolute joy to visit. You slide into the hole and commence a gradual descent along a comfortably sized clean-washed passage. Route finding is straightforward, descending a ladder pitch – The Twenty Foot Pot – en route to Sump 1. Think of those first bold characters who ventured here with their feeble lights, old clothes and felt hats, skittering along in their hobnailed boots.

You will reach the infamous sump in perhaps half an hour, a journey that took the early explorers many hours. This was the site of the first successful passing of a flooded passage back in 1936 and today it is routinely free-dived, being less than a metre in length. For many, reaching this point is enough, but for those keen to see more there are challenging circuits, the most popular of which is the Short Round Trip. This takes a high-level deviation away from the streamway and rejoins the main passage between Sump 1 and Sump 2. Here the caver turns upstream to free-dive out.

Swildon's Hole is the most popular cave of the Mendip Hills and, at nine kilometres, also the longest. It has a vertical range of 167 metres. Park on the village green at Priddy and follow the well-trodden footpath to a blockhouse marking the entrance to the cave.

This is a superb trip, and you will emerge back on the surface with a great feeling of satisfaction and, by way of a bonus, nicely clean, washed kit.

Top right Cascade climb near the Forty Foot Pot.
Right Surfacing at Sump 1.
Opposite The Twenty Foot Pot from the bottom.

HIDDEN REALMS

Wookey Hole

Wookey Hole lies two kilometres north-west of Wells and is renowned throughout the UK as a show cave. Beyond its few hundred metres of tourist walkways is the best part of four kilometres of further passages, much of which are flooded.

Wookey was to be the site of the very first British cave dive, using Standard Equipment, back in 1935. An intrepid duo stepped into the water clad in lead-weighted boots and heavy brass helmets. Their air was pumped by a surface team along snaking hoses that the divers dragged behind them. Since then the cave has continued to host divers, with their increasingly refined equipment, as they have explored through many flooded sections, finally reaching the current terminus at a depth of ninety metres.

Dry sections of Wookey also have a long and colourful history, with archaeological remains being discovered in the first few chambers and an intriguing legend of a witch who once inhabited the cave. In 1975 the tourist area was extended by mining a tunnel and in 2018 a further section was artificially driven to reach another area until then restricted to divers (Chamber 20). From here, in 2020, dedicated explorers finally gained access to the divers' dry extensions in Chamber 24. Today, cavers can undertake a superbly sporting trip involving squeezes, climbing and swimming to see the white foaming water of the River Axe thundering down the cascades in Chamber 24, a sight which is quite awesome. Beyond here, the darkness still beckons.

Above Diver in Chamber 19. **Right** Chamber 1 boat crossing.

MENDIPS AND SOUTHERN ENGLAND

Upper Flood Swallet

Upper Flood Swallet is the longest cave in the Blackmoor Valley, at the head of Velvet Bottom. Currently four kilometres of passages are known with a vertical range of 129 metres. Water from here feeds Cheddar Risings, some five kilometres away and 210 metres below.

Soon after entering the 'Old Cave', a hands-and-knees low section drops into a well-decorated streamway and here the visitor must continue to crawl to avoid damage to the prolific formations. Just after the stream disappears, a damp U-bend – known as the Lavatory Trap – leads on to a boulder choke. This marked the end of the cave until it was finally passed in 2006.

Twists, turns, scrambles and squeezes feel like a marathon. Eventually emerging from the boulders comes as a massive relief and you pass into the large void of the Departure Lounge. A most impressive cave lies ahead.

Malcolm's Way is an easy-going stream passage, well decorated along its length with white flowstone, roof pendants, curtains and straws. Then you reach Royal Icing Junction, the preliminary to Neverland. From here, maintaining exemplary cave conservation is paramount and cavers remove their muddied gear before entering this special place. Stunning is way too much of an understatement. The pristine flows of calcite are just pure white, curtains and stalactites abound, and the strange Pork Pies are unique; this place is simply gloriously magnificent.

Opposite Neverland. **Above** Neverland and the Pork Pies.

HIDDEN REALMS

GB Cave

GB Cave is located about five kilometres north-west of Priddy in the GB Gruffy Nature Reserve. It has two kilometres of passages and reaches a depth of 137 metres. This major cave was first explored in 1939.

GB is a cave on a grand scale. Visitors will find their descent, through a crawl and downclimbs, leads quickly to the streamway called the Gorge, a spectacularly large feature and one of the most memorable in the Mendips. The roof towers above and in the distance a dark tunnel leads on towards the Bridge. Beyond lies the Main Chamber where things become extremely impressive: the cavern measures twenty metres wide by twenty-three metres high. The sense of void here is outstanding by any British standards, and with good lighting it will present stunning vistas of stalactites, flowstone and even helictites if you care to look closely.

Here, the small stream disappears into a chaos of boulders, but around the edge of the chamber several passages can be found. One accesses the Loop Route which presents a fine viewpoint out over the Main Chamber. A second passage high up the wall, best accessed with a ladder, leads to a particularly fine extension and the profusely decorated Bat Passage.

GB is an easy-going, absolutely superb place and it is fascinating to think that Charterhouse Cave passes just tens of metres beneath.

Above Bat Passage.   **Right** White Passage.

HIDDEN REALMS

Charterhouse Cave

Left The Blades.
Above The Frozen Cascade.

MENDIPS AND SOUTHERN ENGLAND

Charterhouse Cave lies very close to GB Cave. This is an exceptionally long, challenging and fine cave, the deepest in the Mendips. Lead miners first discovered natural cave passages here hundreds of years ago, but it wasn't until 1972 that the current entrance was opened by members of the Sidcot School Speleological Society.

The place starts small, with an interesting tight squeeze just a short distance from the surface. This is a good gauge as to the challenges that might be expected ahead. The obstacle course continues until eventually the cave opens out in spectacular style at the Citadel. This large chamber was discovered in 1982, but a massive boulder blockage just beyond – Chill Out Choke – effectively halted progress towards the water's outflow at Cheddar for many years. In 2008 a momentously dramatic breakthrough was made, taking the cave rapidly deeper. Further discoveries in the following two years more than doubled the length of the cave. Today, the system extends to almost five kilometres with a vertical range of 228 metres.

Charterhouse certainly has sporting passage and superb formations. Anyone passing the infamous Chill Out Choke will be impressed by the dedication of the diggers, both in opening up and in stabilising this arduous and vertical boulder ruckle. Thereafter follows the Narrows ... beware, the name hints at the difficulty! Any trip beyond the confines of Portal Pool is a serious operation and highly weather dependent. Dry weather is essential for a visit to the furthermost reaches of this great system.

HIDDEN REALMS

Reservoir Hole

Reservoir Hole is a discrete cave located in the famous Cheddar Gorge. An inconspicuous gated entrance belies the wonders hidden inside. It took many years and extreme dedication on the part of several generations of cavers before any significant discovery was made. Then, in 2012, the lengthy excavational project achieved a startling success. A ten-metre pitch was descended, and the most incredible sight appeared. The team of locally based explorers found themselves staring into a *big* black void, unquestionably the largest cavern beneath the Mendip Hills and arguably the UK. Moreover, it was magnificently decorated ... its name today is The Frozen Deep.

The Frozen Deep is conservatively in excess of sixty metres long and forty metres wide. It presents three centrally positioned, glorious, glistening, white formations, while off to the sides lie other arrays of towering flowstone. To say that this is a magnificent cavern is a gross understatement. With a backdrop of silence and blackness, the setting is stupendous. There are other passages in the cave, both above and below the grand cavern, with possible leads that, in the fullness of time, will yield further discoveries. But venturing to The Frozen Deep is enough. It is such a privilege to visit this place – one of the crown jewels of British caving.

Left The largest pillar in The Frozen Deep.
Top The immense cavern of The Frozen Deep.
Above A small grotto within The Frozen Deep.

HIDDEN REALMS

Shatter Cave

Two kilometres east of the village of Oakhill is Fairy Cave Quarry where Shatter Cave is found. It was discovered during quarrying operations in 1969 and, like the neighbouring caves of Withyhill and Fernhill, was originally part of one altogether more expansive complex. The magnificent calcite decoration in these caves is in contrast with others in the quarry, which present interesting trips but lack the beautiful speleothems.

After crawling through a concrete entry pipe, visitors to Shatter Cave are quickly introduced to an area of passage that has suffered evident blast damage from the working era. Amazingly, Canopy Chamber, with its large flowstone formations, remains intact despite its proximity to the quarry face. Moving beyond this takes you deeper into the limestone and a series of yet more memorable chambers and lovely grottoes.

You will remember Tor Chamber for its stalagmite boss shaped like Glastonbury Tor, the appropriate naming of the Leaning Tower of Pisa and the translucent beauty of the pristine white Angel Wing curtain. Pillar Chamber then makes a spectacular climax to any trip. With its delicate crystalline floor and a two-metre-high column, you will recall this sight for ever more. Shatter may not be a long cave (it is 1,250 metres in length) or particularly challenging, but it surely presents British caving in a spectacular light.

Above left Tor Chamber.   **Above right** Canopy Chamber.   **Right** Pillar Chamber.

HIDDEN REALMS

Withyhill Cave

Withyhill, like Shatter Cave, is located in the abandoned Fairy Cave Quarry, the site of major industrial activities from the 1920s to the 1970s. It may only be 1,460 metres in length with little or no height change, but wow – this is a wonderful place.

The entrance consists of a concrete pipe resembling a road culvert rather than a cave. But no sooner are you through the portal than you start to appreciate how special a visit here is. Few caves possess such a density of calcite. It is profusely decorated, and many formations are vulnerable to damage, however carefully the small parties move through.

Grotto follows grotto; there are so many flowstone structures and exquisite helictite growths that you really will want to take your time. Marvel at the Elephant's Trunk, Column Chamber and Helictite Corner. Perhaps the crowning glory is to be found in the further reaches of West Passage where a squeeze up on the left-hand side reveals the charming Green Lake Chamber. This is a magical little oasis of crystal-clear water, surrounded and overhung by a closely packed array of formations. The beautiful green pool is a most memorable sight for careful parties who journey to this point. You will stand there in silence. Having seen it, you know that this place is truly special.

Above left Column Chamber.   **Above right** The Elephant's Trunk.
Left Helictites.   **Right** Green Lake Chamber.

HIDDEN REALMS

Stoke Lane Slocker

Stoke Lane Slocker lies just north of the village of Stoke St Michael in the eastern Mendips. The system extends to over two kilometres, but a significant amount lies beyond a series of sumps. With its low entrance crawl, Stoke Lane begins in a less-than-inspiring manner. There now follows hands-and-knees grovelling which continues rather longer than one might wish. Beyond the Nutmeg Grater constriction, it looks like ten metres of flooded rift spells the end. This small, cold and gloomy pool is Sump 1.

While people routinely plunge through Sump 1 in Swildon's Hole, this one is much less frequented. It's only just over half a metre long but somehow feels more committing. Passing inward is relatively straightforward as you simply dip under an arch below the left-hand wall of the canal. The return needs to be more careful to ensure you don't overshoot.

Providing the weather is dry and stable, take the plunge and you will be utterly amazed by what lies beyond. Within a short distance the place undergoes a complete – and welcome – transformation. The stream passage gains size and requires you to scramble over, between and up through boulders. Then you encounter a substantial foam-ridden pool, Sump 2, the end of the cave for a non-diver.

But look behind you and you will see a big, dry tunnel above. Clamber up into the Throne Room where some fabulous formations will be unveiled – just reward for your efforts and a glorious end to this underground adventure.

Left Throne Room grotto.   **Above** The King in the Throne Room.

HIDDEN REALMS

Pen Park Hole

Pen Park Hole is unique amongst British caves, situated as it is beneath Southmead housing estate in Bristol. Virtually the entire route, from the entrance to the spectacular main cavern, is encrusted with dogtooth spar crystals. These provide strong evidence that this is one of the few UK sites formed by rising hydrothermal waters. Nowadays, the level of the lake fluctuates by as much as twenty-seven metres, but the deep-seated, warm upwelling has been cut off leaving it at just a cool ten degrees. Here is found our only cavernicolous population of the tiny shrimp-like crustacean, *Niphargus kochianus*.

The place has a fascinating history: in 1669 Captain Samuel Sturmy made the first complete descent of the site along with an accompanying miner. It is reported that Sturmy developed a headache soon after surfacing, then a fever, and just four days later he died. Amazingly, another fatality occurred here in 1775 when the local vicar tried to plumb the shaft. A branch he was holding broke, and he fell to his death. Less dramatic – but certainly important – mapping was undertaken in 1682 and the following year this became one of the earliest cave surveys to be published anywhere in the world.

Although only short – the cave is around 340 metres in length and sixty metres deep – its unusual features make Pen Park Hole an extraordinary and very different place to visit.

Above Crystal encrustation in the entrance passage.
Right The lake.

HIDDEN REALMS

Bath Stone Mines

The many mines located in Wiltshire between Corsham and Bath are known collectively as Bath Stone Mines. Their sheer extent is incredible and route finding in some sites is a real challenge. To try and understand their sheer length and complexity, it is worth noting that Box Mine alone contains ninety-five kilometres of tunnels.

The historic origins of the underground workings are unclear, but we know that the Romans and every age thereafter has understood and valued the quality of the stone. Blocks hewn from these tunnels adorn the beautiful facades in Bath, and many other towns and cities across the country. A vast quantity of limestone has been laboriously removed – by hand, not using explosives – over the past 2,000 years.

All manner of artefacts are present, including stone blocks, spades, chisels and rock-cutting saws, as well as historical wall paintings and tally charts. Perhaps the finest features are the old wooden cranes which are a magnificent sight to behold.

The showpiece of Box Mine is its largest chamber, the Cathedral, where a faint patch of daylight permeates from a surface opening. It measures approximately fifty-seven metres in length, nine metres wide and thirty metres high. This mine is a splendid place but countless people have become disorientated here. Confronted with head-scratching confusion, many have inscribed coloured markings on the walls in an attempt to aid navigation. Unfortunately, this does not always achieve the desired result. So, if you visit, ensure you know your way back out and *don't deface the walls*.

Above left Crane in Swan Mine. **Above right** Horse trough in Swan Mine. **Right** The Cathedral in Box Mine.

HIDDEN REALMS

Emmer Green Chalk Mine

The chalk mines of Southern England are not that long or deep, but provide an interesting little venture with equal historical fascination. One such is Emmer Green Chalk Mine, with its origins possibly as early as the seventeenth century. This lies in a quiet corner of a garden at the northern edge of Reading, Berkshire.

Chalk, with associated flint deposits, has been mined for many hundreds of years in this area. In the past, the Reading district was renowned for its brickmaking industries. This is significant due to the simple fact that a percentage of chalk was added to the brick clays before firing, to counteract shrinkage. Apart from building purposes, chalk was also a valuable product in agriculture and used to lime fields.

The mine below Emmer Green appears to have been associated with the earliest industrial use of the land hereabouts. At some time in the nineteenth century the site was abandoned, entry was blocked and the mine forgotten. It was not until a narrow shaft was discovered in 1977 that its existence came to light.

Less than twenty metres below ground you drop into a totally different world, where everything is a creamy off-white with walls presenting a distinctly crumbly looking appearance. It is evident that the tunnels have been laboriously hand-picked, leaving a beautiful curved cross-sectional shape. Passageways are generally of the order of five metres high and up to three or four metres wide with some sizeable chambers. The place is totally dry and, keeping a careful eye, artefacts together with fascinating graffiti, dating back at least to 1776, will be found.

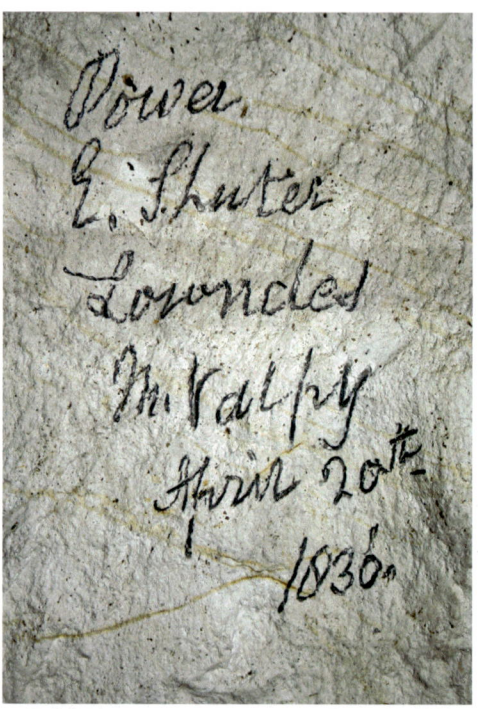

Top Inscriptions on the walls. **Above** Artefacts found in the mine. **Right** Chamber within the mine.

HIDDEN REALMS

Pridhamsleigh Cavern

There is a limited amount of limestone in Devon and relatively few caves, but near the village of Buckfastleigh lie a couple of notable systems of speleological interest, namely Pridhamsleigh Cavern and the Baker's Pit–Reed's Cave system.

Pridhamsleigh Cavern, or 'Prid', has a surveyed length of 1,100 metres. This consists primarily of a fascinating maze of fossil phreatic passageways. Venturing inside the impressively large entrance, you will follow in the footsteps of many who gained their first experience of the subterranean world here.

Pridhamsleigh presents what appears to be very challenging navigation to a 'first timer', but, if you follow your nose by taking the easiest options, you will traverse the cave without too much worry. As in any underground complex you do need to note significant landmarks and step carefully as the passages are all pretty well travelled with polished rock and a slight covering of slippery mud.

There are flowstone formations in the cave, but sadly many have suffered through the actions of thoughtless explorers in the past. Visitors will be prudent to avoid areas of water on the visit, especially the twenty-metre-long terminal Lake, as this is thirty metres deep!

Prid is an easy cave by any standards with a lovely charm of its own. It's literally a stone's throw from the A38 and just a short walk from the parking spot.

Above Flowstone grotto.   **Right** The Lake.

Reed's Cave

Reed's Cave is found in a long-abandoned quarry just a few metres from the Pengelly Centre in Buckfastleigh. It is linked to Baker's Pit which is the longer part of the underground network, with some two and a half kilometres of once-well-decorated passages. Unfortunately, the site suffered considerably from the ravages of souvenir hunters in the early twentieth century. Reed's, by comparison, was not as readily accessible and the main passages were not discovered until 1939. By this time there was, thankfully, greater appreciation for the underground environment. Today, the connection between Reed's Cave and Baker's Pit has been blocked ensuring that the former cave remains conserved.

Reed's Cave begins by scaling a stepladder in the impressively large entrance portal to access the gate. There follows a crawl, then an excavated squeeze, and quickly one reaches the substantial Easter Chamber and some very fine curtains. At the opposite side of the chamber lie a couple of small tunnels along which will be found the unforgettable sighting for which Reed's is renowned: the Little Man formation. As the name hints, this calcite structure resembles a small man with arms outstretched and sporting a top hat. The legend of seventeenth-century local, Squire Richard Cabell, who was described as a 'monstrously evil man', is directly associated with this formation and the story makes for wonderful reading.

Reed's may only be 900 metres long but it's a great insight into subterranean Devon.

MENDIPS AND SOUTHERN ENGLAND

Opposite The Little Man. **Above** Easter Chamber.

HIDDEN REALMS

Tywarnhayle Copper Mine

Cornwall and South Devon are famed for the production of copper and tin. Both metals are scarce but have been found and worked in the region for 4,000 years. By the eighteenth century Cornwall had become the world's major producer of these metals and by the early nineteenth century this was the most technologically advanced mining district on earth. In the peak years around 1870 it is estimated that one quarter of the Cornish population was employed within the industry.

Tywarnhayle Copper Mine lies on the outskirts of Porthtowan in North Cornwall. At the outset of activities in 1750 it was known as Wheal Rock Mine, later acquiring its current name. After 1907, at the latter stage of its working life, the site became London's Royal School of Mines training mine, specialising in surveying and timber preservation.

There are several entrances to this underground complex including one impressive shaft (James' Shaft), in excess of sixty metres deep. Below ground, a variety of accessible tunnels spread out beneath the hill, extending to water level. As with all mines, some areas are unstable, but there are some really interesting features. The ladderway and incline shaft are very impressive, and looking more closely there are enchanting, colourful deposits derived from the host rock. Veins of ore are not obvious, but small areas are wonderfully decorated with blues and greens. Lustrous, bright pink patches evidence pure copper metal, while the tiny yellow accumulations of sulphur are quite special. The growth of more substantial, dark, iron-like flowstone formations testify to the rapidity of such development since the tunnels were driven.

Left Incline shaft. **Above** Engine house.

HIDDEN REALMS

Rosevale Tin Mine

Rosevale is a mine with a difference. This relatively small network, near Zennor in North Cornwall, is typical of the small Cornish tin mines which existed during the Industrial Revolution. Many of these were less than 100 metres deep and employed only a handful of people.

The precise beginning of metal extraction in this area is uncertain, but evidence of early working can be found on local moors in the form of granite mortars and rounded stone pounders. These were used in the Bronze Age to crush the rock to enable the recovery of the heavy metallic ore. By the time Rosevale mine was in operation in the nineteenth century, the processes of ore extraction had become mechanised. Like many such operations, the place had a chequered history, finally ceasing work in 1914. The site was never one of Cornwall's great mines, but it does remain one of the only complete examples of underground tin workings in the area.

There are three distinct interconnecting levels in this mine, the longest of which extends almost 300 metres into the hillside. From the entry point at the Level 2 portal, a fascinating insight to the historic enterprise is quietly revealed. Following a vein along the adits, you will see timbered stopes, ladders, ore chutes and a wheelbarrow – indeed all manner of industrial artefacts once common in the region.

For over forty years a group of dedicated volunteers has been working on a conservation project at Rosevale, aiming to restore the mine and preserve the area's important heritage. This restoration is an impressive industrial archaeological project.

Above left Timber supports. **Above right** The entrance to Level 2. **Right** Kibble, tram and ore chute on Level 1.

FOREST OF DEAN

Between the rivers Wye and Severn lies the undulating upland of the Forest of Dean, an area long famed for its iron mining. Much of the industrial evidence is now disguised amid a glorious expanse of fine deciduous woodland, but wander into this forest and, at various places, areas of old workings can still be found. Labyrinths such as the tourist attraction of Clearwell Caves are exceptionally complex and extensive.

Secreted a short distance to the north and west are especially interesting caves. Longest of these is Slaughter Stream Cave, with its rich variety of archaeological remains, but on the banks of the River Wye lies Otter Hole – wholly unique in British caving and fabulously decorated.

48 Clearwell Caves
49 Slaughter Stream Cave
50 Otter Hole

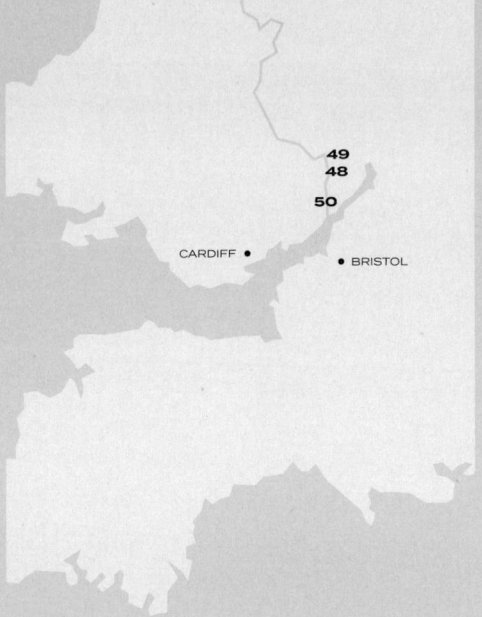

Left Noxon Park Iron Mine, near Clearwell Caves.

HIDDEN REALMS

Clearwell Caves

The iron mines of the Forest of Dean are astonishing. The sheer extent of the underground workings is difficult to get your head around, even for a seasoned underground explorer. Ore has been extracted here for thousands of years.

One fine example of these forest sites is Clearwell Caves, part of which runs today as a show mine. For an informed educational experience this is most certainly the place to visit. The mine is part of a thirty-kilometre complex extending in a maze-like fashion from the village of Clearwell to Coleford, a couple of kilometres distant. It has a depth of 180 metres, of which the lower fifty metres are now flooded. Iron mining here ended in 1945 but small-scale extraction of the various pigments of ochre still continues today.

The show mine is open all year but at Christmas the decorations set out around the tourist tunnels are world class. For a 'wild tour' into the deeper parts of the mine you will be led by a knowledgeable and professional guide. You will appreciate not stumbling around using a tallow candle, which was the most common option for the original miners, but it may be interesting to see how long it takes to remove the reddish ochre staining which colours hands, clothes and footwear. It's said that you can always tell a Forest of Dean underground explorer just by looking at their wellies!

Left The show mine at Christmas.
Above Old Ham Mine, part of the Clearwell Caves complex.

HIDDEN REALMS

Slaughter Stream Cave

Slaughter Stream Cave lies three kilometres north of Coleford, a short distance from the Wye Valley. When you walk towards the cave from the parking area, it feels far from being a caving venue. Little do people generally realise that beneath their feet lies one of the most extensive networks in Southern England. At over thirteen kilometres long, a good trip in this oft-overlooked system is possible in most weather conditions. The depth range is 130 metres and, with the resurgence three kilometres to the west, there is undoubtedly a lot more cave to be revealed here.

Excavation at the Wet Sink entrance started in the 1950s. With periods of activity and then breaks, it was not until 1991 that local cavers made the big discoveries. Once through this arduous dig, the explorers entered some seven kilometres of passages without needing to move another rock! Fixed ladders start the early part of the descent, followed by two SRT pitches, the deepest of which is twelve metres.

From here, many people turn downstream. With water depth only occasionally over the knee, this is a pleasant walk through sculpted passageway, but all too soon the stream disappears into a sump. It's the fossil passages, upstream and above, that provide the long excursions in this cave. Terrain is varied: dry sandy crawls, rose-coloured limestone, boulder floors and some very unusual geological features. For the observant, there is much to see here, not least the intriguing remains of 'Norman' the dog – lying on a sandy floor, far, far from the entrance we use today!

Left The Chunnel. **Above** Norman the dog.

Otter Hole

Otter Hole is very special and it's a place that should feature right at the top of any caver's aspirational trip list. The cave, near St Arvans in the Wye Valley, is entered very close to river level – so close in fact that the entrance series of this three-and-a-half-kilometre system is dramatically affected by the rise and fall of the tide in the Severn Estuary.

This is certainly a trip that you will muse long and hard about. Where normally one can choose casually when to kit up and get going, here it's imperative that you are thoroughly prepared and totally time aware. The tidal sump opens and closes twice every day, and you really cannot afford to disregard or fall foul of the timings: the speed at which the water level rises is, from personal experience, alarming. This place demands respect.

What a treat awaits you beyond the muck and slurry of the squalid entrance passages. You ascend into a fossil section and the environment is totally transformed. Here, deep beneath Chepstow Racecourse, the place is stupendously well decorated. There are calcite formations at virtually every turn; conservation is of the utmost importance. The first of the great galleries is the Hall of Thirty, named after the sheer number of stalagmites adorning the floor. And so it goes on ...

You are guaranteed to be blown away by the sights. This is the most beautiful cave in the British Isles.

FOREST OF DEAN

Left The tidal sump. **Above** Hall of Thirty – spot the two cavers!

PEAK DISTRICT AND CENTRAL ENGLAND

The Peak District is a wonderous world for the underground enthusiast where a labyrinth of mine and natural cave passages are often inextricably woven together. Set at the foot of Mam Tor hill lies Castleton and here the gaping cavernous entrance of Peak Cavern beckons loudly. Over the centuries, this locality has been widely prospected for lead and the semi-precious mineral Blue John. The places the early miners ventured are breathtaking; they explored with tallow candles and scaled unbelievable vertical shafts. Many of their passages have been lost to the mists of time and present-day enthusiasts are still working to follow in the footsteps of those courageous men.

Beyond here, Central England has still more to offer including some fascinating mines which certainly deserve their inclusion in this book.

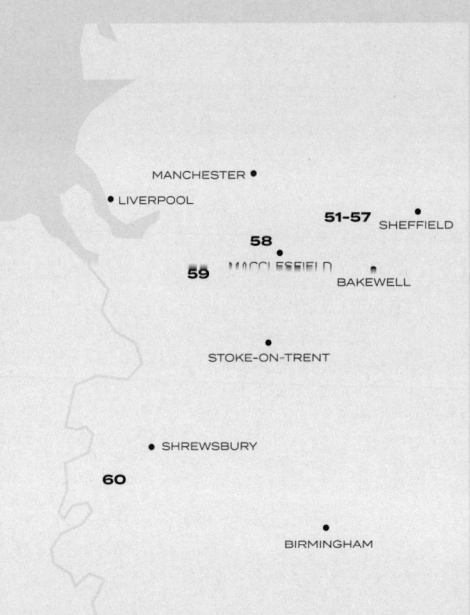

51	Peak Cavern
52	Speedwell Cavern
53	Titan Shaft
54	Oxlow Caverns
55	Giant's Hole
56	P8
57	Blue John Cavern
58	Alderley Edge Mines
59	Winsford Rock Salt Mine
60	Snailbeach Mine

Left Winnats Pass above Castleton.

HIDDEN REALMS

Peak Cavern

Above Squaws Junction. **Opposite** Phreatic tunnel downstream from the Surprise View Pitch.

PEAK DISTRICT AND CENTRAL ENGLAND

Peak Cavern is a spectacular show cave in the village of Castleton in the Derbyshire Peak District. With its connections to places such as Speedwell Cavern, James Hall's Over Engine Mine and Titan, the overall complex stretches to approximately seventeen kilometres with a depth range of 235 metres. Hydrologically, it is the outlet for places much further afield such as P8 and Giant's Hole.

The entrance, which is the largest of any cave in the British Isles, is always worth a stroll to see even if you are not going underground. For a caver, a subterranean visit is something never to be forgotten as you access a superb system that will guarantee to lure you back for more challenging trips. It's easy to understand how this highly impressive place generated so much interest with adventurous types way back in the earliest of times. For example, in 1777 an unnamed gentleman attempted to free-dive through the Buxton Water rising within the cave but unfortunately knocked himself out on the rock and had to be dragged back to the edge with some difficulty!

The exceptionally large entrance tunnel ends all too soon at the Buxton Water rising but an aqueous bypass leads through some chest-deep wallows to regain the enormous main passage at the six-metre-high Surprise View Pitch. Downstream leads through the most wonderfully oval phreatic tunnel in the UK to reach the other side of the sump you saw early on. Upstream from the pitch is just a fabulous romp, ultimately reaching an abrupt end at the Far Sump. Along your route you will pass towering avens and side tunnels, such as those to Moss Chamber, Ink Sump and Treasury Sump. Peak Cavern is a real classic.

HIDDEN REALMS

Speedwell Cavern

This site exudes a very special character and has featured prominently in the annals of lead mining history with activities stretching back well over 200 years. Today, Speedwell Cavern is a popular show cave complex located near Castleton, at the foot of Winnats Pass. The casual sightseeing tourist will buy a ticket and be transported deep into the hillside on an electrically powered boat. It's a unique trip for England and the guide will impart wonderous tales along the way.

For the caver, the passages that lie beyond public limits are both interesting and challenging. An active waterway, indeed a small river by caving standards, flows through the further reaches of Speedwell and this is one of the most sporting stream passages in the UK. You will find deep sumps at the upstream terminus, one of which – Whirlpool Rising – is quite unique as it periodically syphons to discharge a veritable tsunami-like wave of water. Imagine flushing the toilet and think what it must be like somewhere further down the pipeline …

The river passage in Speedwell was first discovered by the lead miners who came in via shafts from other entrances such as James Hall's Over Engine Mine high above. Some of those intrepid characters left fascinating evidence of their visits plain to see on the cave walls. With its connections to a number of other underground sites, this is a fascinating and important place.

PEAK DISTRICT AND CENTRAL ENGLAND

Left The Main Rising. **Above** Lower Bung Streamway cascade.

Titan Shaft

Titan is in a class of its own. This site, which lies high on the Hurd Low hillside above Castleton, is breathtaking. Here lies a fascinating back story of extreme dedication, determination, skill and a heap of 'derring-do'.

The tale begins more than 200 years ago with lead mining and tunnels that were subsequently 'lost'. Fast forward to 1981 when lengthy passages and evidence of miners' activity were found by diving through Far Sump in Peak Cavern. Spurred on by this, local cavers were prompted into a period of intense activity. Further discoveries were made, and on New Year's Day 1999 Dave 'Moose' Nixon and team emerged at the base of an immense vertical shaft that defied the strongest lights – a shaft which they subsequently named Titan.

What followed was a climbing adventure. Eventually, after an audacious ascent, Titan was scaled to a height of 143 metres and established as the longest free-fall natural shaft in the UK. The challenge then was to achieve an upper entrance ... and so another epic excavational project began.

By 2003 an infilled mine shaft had been discovered and cleared. This dropped into a short horizontal passage to emerge 120 metres above the floor of the giant Titan Shaft. Today, the so-called Trade Route descent involves a forty-five-metre entrance pitch followed by two drops down the Titan Shaft, split by a welcome break at the Event Horizon. To descend (or ascend) this spectacular void is the ultimate SRT undertaking in the British Isles.

Above Explorer and pioneer Dave 'Moose' Nixon emerging from the entrance. **Right** The main shaft.

HIDDEN REALMS

Oxlow Caverns

PEAK DISTRICT AND CENTRAL ENGLAND

Oxlow itself is a delightful, pretty much dry, sporting trip involving a number of pitches to reach the bottom. All of these are relatively short and straightforward, with good secure belay points, making this a popular venue for SRT novices to practise their skills. At the bottom of the third pitch is a vantage point giving a view into the impressively proportioned East Chamber which can be accessed from here. Continuing down westwards, a steep handline descent leads to the head of the fourth pitch. Below lies the West Antechamber, with a low arch leading to another cavernous void, one of the largest in Britain.

Located close to the head of Winnats Pass, Oxlow is connected with Maskhill Mine which lies nearby on the hillside. The combined system yields an overall length of 1,600 metres and offers potential for trips to depths of 145 metres. Maskhill Mine must be treated with much greater care owing to areas of loose rock and the potential to shower this on anyone below. Connection from Oxlow can also be made with Giant's Hole via the aptly named Chamber of Horrors. This constricted duck-cum-sump is definitely not for the faint-hearted and is rarely attempted – it's an extreme proposition.

Left West Chamber.
Above left Ascent of the entrance shaft.
Above right Ascent of the fourth pitch.

HIDDEN REALMS

Giant's Hole

In the early 1950s determined excavations were made in the large entrance sink of Giant's Hole, found high on Rushup Edge. Success revealed open cave and the route onwards to Garlands Pot. The drop here is only about six metres and leads directly into the Crabwalk, an awkward section of passage in a high and narrow, twisting fissure. After 400 metres of sideways wriggling and shuffling there are a couple of little climbs, the second of which is the Comic Act Cascade. Given the fine spout of water shooting over the smooth rock, one can readily understand why a good solid metal ladder has been installed. Beyond, the Crabwalk continues for a short distance, increasing in size and becoming well-sculpted in shape, until a sump blocks the stream.

Turning right shortly before this yields a climb to a higher level and now you are homeward bound. Passing through the wet Giant's Windpipe, you eventually complete your circuit by a fourteen-metre descent back down to the Crabwalk … a short distance before Garlands Pot. With just over three kilometres of passageways, this is one of the most popular caves in the area and it's always a good trip.

Above Garlands Pot. **Right** End of the Crabwalk.

PEAK DISTRICT AND CENTRAL ENGLAND

P8

P8, also known as Jackpot, is located in a line of swallets near Perryfoot, on Rushup Edge, approximately four kilometres from Castleton. It is fabled for its wet and sporting descent, and is very popular with cavers.

While the overall extent of the cave is about 1,800 metres, it is worth noting that approximately 600 metres of this lies beyond the limit accessible to non-divers. The water from the system drains to Speedwell Cavern but it seems improbable that any passable connection between the two points will ever be achieved.

The entrance lies an easy kilometre's walk from the car parking area and was first explored in 1964 by members of the British Speleological Association. From the outset, as you climb down with the water, the aqueous nature of this cave asserts its presence, and the route is followed along a narrow, clean-washed streamway. Small cascades are easily overcome, although the two-and-a-half-metre Idiot's Leap requires a little more care. Some eighty metres from the surface comes the First Pitch and at the bottom the visitor is presented with a choice of routes. Either follow the streamway directly to the Second Pitch or climb up into the high-level passages to descend an alternative dry pitch. There follows a fine spacious section with water chutes, pools, flowstone cascade and easy walking along T'Owd Man's Rift to a froth-ridden sump. With a total depth of seventy metres, this little adventure will take several hours and you should emerge feeling pleasantly invigorated and with kit completely clean.

Top right First Pitch.
Right Second Pitch (alternative descent).
Opposite First Pitch from the bottom.

HIDDEN REALMS

Blue John Cavern

Above A small vein of Blue John.　**Opposite** End of the show cave.

Blue John Cavern was originally discovered by prospecting lead miners over 250 years ago. Today, the site is well known as a tourist venue, lying just a couple of kilometres from Castleton on the flank of Mam Tor. The complex as a whole extends to 1,274 metres of passageways with a depth of 90 metres.

In their excavations the miners revealed substantial natural cavities and also Blue John, a unique and subsequently highly sought-after semi-precious mineral. Occasional tourists were taken into the caverns from about 1770, when visitors had to descend via a crude miners' stemple stairway. Access was considerably improved by 1840 with the creation of a flight of stone steps. Blue John is still worked periodically to be crafted into beautiful jewellery, renowned in gift shops throughout the area.

Anyone venturing underground here will be impressed by the sheer size of the natural cavities. Places like Crystallised Cavern, Lord Mulgrave's Dining Room and the Variegated Cavern are all twenty-seven metres or more in height. Crystallised Cavern has yet one more important claim to fame: this was the site of the world's very first underground photograph, taken by Alfred Brothers on 27 January 1865.

Photography, in any environment, in the 1860s was not common. Cameras were heavy, a tripod was essential, and the images had to be made on glass plates. The exposure of each image took five minutes and was made possible by burning magnesium to generate sufficient lighting. The first commercial production of magnesium in the world was begun in Manchester in 1863 and the proximity of Blue John Cavern to the source of manufacture was clearly a major factor in Brothers' selection of the venue for his photographic experimentation.

HIDDEN REALMS

Alderley Edge Mines

Alderley Edge Mines are located between Macclesfield and Manchester. This is an intriguing complex of more than seven mines scattered across an area of several square kilometres. A variety of minerals, including copper, lead and cobalt, have been mined over the years and the earliest workings may well have dated back to 1900 BC, during the Bronze Age. The most productive phase was during the latter half of the nineteenth century when all the mines were connected at depth by the Hough Level. Work ceased here in the 1920s.

Above left Fixed ladder and copper flow. **Above right** Coffin level.
Right Large chamber at the bottom of West Mine.

There is a surprising amount of tunnel to be explored across the network, and just two of the sites, Wood Mine and West Mine, contain over twelve kilometres of passageways. Seeing the small hand-picked 'coffin levels' at Engine Vein Mine is thought provoking, while, in complete contrast, the tunnels in West Mine are surprisingly large, dry and sandy. Geologically, with its sandstone rock formation, this is a really fascinating place and an interesting diversion from the normal subterranean scene in the Peak District. The influence of faulting is clear to see at various points, and, with their veins of mineralisation, these mines are certainly a very colourful curiosity.

Taking time to explore, you will see impressively cut shafts and all manner of artefacts that may be found throughout the complex. Roman coins have been recovered and close by an intriguing rock carving can be seen in the wall. Is this possibly a Roman shrine?

Winsford Rock Salt Mine

Winsford Rock Salt Mine in Cheshire is a rare example of a working mine of national importance. It is remarkably unknown by most underground enthusiasts, and due to issues of security this is not an easy site to access. Wow, this is an awe-inspiring place!

Salt had been extracted in the locality since Roman times, but it was not until 1844 that Winsford Rock Salt Mine was opened. Today this thirty-one-square-kilometre site is Britain's largest supplier of rock salt – used to thaw icy roads during winter. Mining and storage of the strategically important mineral is undertaken using sophisticated technology and anyone privileged to visit will be highly impressed by both the scale and safety procedures in the complex. Gone are the days of drilling and the use of explosives. Huge, alien-like cutting machines – continuous miners – excavate the 220-million-year-old salt deposits at depths approaching 300 metres below the surface. Substantial pillars are left untouched to support the roof and, standing next to these in the vastness of the maze-like network, one feels distinctly small.

In 1998 a secure storage facility was established here utilising the empty voids left behind after mining. Today, 150 metres below ground, are found all manner of important items. What a far cry this place is compared to the mines of the nineteenth century.

Below left Continuous miner and operator.
Below right Salt encrustations surround an old mine shaft.
Right The end of the conveyor.

HIDDEN REALMS

Snailbeach Mine

Snailbeach Mine lies in the small village of Snailbeach, fifteen kilometres south-west of Shrewsbury. In its heyday it was the most important lead mine in Shropshire.

Lead was worked in the locality in Roman times, but the earliest recorded mention of the mine was in 1676 when miners from Derbyshire started working here. A boost in fortunes was seen in 1881 when a compressor house and chimney were built to provide power for pneumatic drills and winches. At its peak 500 people were employed and it is reputed to have yielded the greatest volume of lead per acre of any mine in Europe. Extraction of lead continued up to 1911 and all underground work had ceased by 1955.

In the early 1990s Shropshire County Council undertook extensive restoration to make some of the shallow workings safe. Today, there is a visitor centre and heritage complex, parts of which are open to the general public. Indeed, Snailbeach is perhaps the most complete collection of mine building remains in England.

Below ground, the route to the bottom is well rigged for SRT and the shafts are nice and dry. As in any such mine, you need to take care not to dislodge rocks in the sloping parts, but all told it's a fine trip. Snailbeach is an impressive place with very substantial caverns and there are interesting sets of relics to be found on various levels. Undoubtedly, underground enthusiasts will be more than pleased with a day visit to this mine.

Above Tram on 40 Yard Level.
Right Barytes vein in the Perkins Level.
Opposite Shaft at the end of Day Level.

YORKSHIRE AND NORTHERN ENGLAND

The Pennine Hills and Yorkshire Dales present a beautiful vista. But on the windswept tops you will quickly see the very 'bones' of the landscape, eroded and fissured as rainwater dissolves the limestone rock.

This is the land of vertical caving. Venture beneath the surface and there are stunning sights – clean-washed, sculpted shafts often spray-lashed by waterfalls that thunder down, carving out the rock. With knowledge and skill, this is a wonderful area to visit.

Explore further afield and you will find the mines of Cumbria. From the neatly crafted tunnels of Nenthead to the blue hues of the copper mines, each has its own character and story to tell.

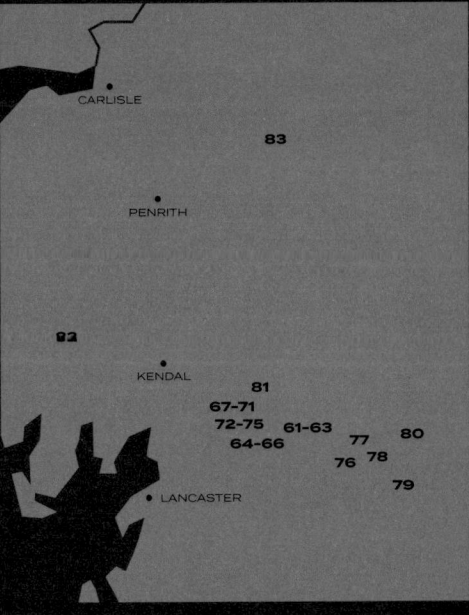

61	Sell Gill Holes
62	Long Churn Caves
63	Alum Pot
64	White Scar Cave
65	Gaping Gill
66	Juniper Gulf
67	Lancaster Hole
68	County Pot
69	Shuttleworth Pot
70	Notts 2
71	Ireby Fell Cavern
72	Simpson's Pot
73	Rowten Pot
74	Kingsdale Master Cave
75	Illusion Pot
76	Sleets Gill Cave
77	Dow Cave
78	Mossdale Caverns
79	Stump Cross Caverns
80	Goyden System
81	Ibbeth Peril Cave
82	Coniston Copper Mines
83	Nenthead Mines

Left Limestone pavement on the flanks of Ingleborough.

HIDDEN REALMS

Sell Gill Holes

Sell Gill Holes is an ever-popular site, a two-kilometre walk north from the village of Horton in Ribblesdale. The pair of entrances, set either side of the Pennine Way, are easy to find. The wet entrance, which hosts the Goblin Route, is to the right, upslope of the track. The more often used Fossil Route lies through the entrance to the left, down a short gully. The site was first bottomed by members of the Yorkshire Ramblers' Club in 1897.

Generations of potholers have cut their teeth on the pitches in this lovely cave, learning and practising ladder, lifeline and SRT techniques. It may only be 500 metres long and possess a modest depth of seventy-eight metres, but it offers plenty of variety and the possibility of rigging both entrances and conducting a crossover trip. The Fossil Route has an abundance of anchors on each of the three pitches and, as these drops are relatively short, communication is straightforward.

The Goblin Route presents rather more of a challenge, especially in anything other than dry conditions. Some fine acrobatics are required to reach the head of the final drop, Goblin Shaft, and the rock sculpting along the way is superb. This route is highly memorable and really sporting in good weather; if the forecast is poor, it's a place to avoid. Both routes unite in the large, dark-walled and impressive Main Chamber, from which point the stream then leaves on its final journey along the meandering passage to the sump.

Top right Wet entrance leading to Goblin Shaft.
Right Top of the second pitch, Fossil Route.
Opposite Ascending Goblin Shaft.

HIDDEN REALMS

Long Churn Caves

Long Churn Caves are located on the eastern flank of Ingleborough, an easy twenty-minute walk up from the car parking area near the hamlet of Selside. Upper Long Churn Cave has a length of 762 metres and a depth of eighteen metres, while the second, lower, part of the system is only 366 metres long.

These caves are one of the most popular sites with novices in the UK. Visitors will certainly not be disappointed with the varied terrain. Entering at Upper Long Churn you quickly encounter a great water chute at Dr Bannister's Handbasin – care is needed in high water conditions. Then some surprisingly easy going along a superbly clean-washed passage leads to another short waterfall dropping into Lower Long Churn.

Here the stream soon turns off as it makes its way towards Diccan Pot. Following the main passage, you negotiate some deep, icy pools and, if so inclined, wriggle through the notorious Cheese Press! After a scramble down, the route leads you to the Dolly Tubs pitch. This is only fourteen metres deep but yields a spectacular vista out and across the magnificent daylight chasm of Alum Pot, one of the finest and most beautiful views you'll see in a British cave.

Above The Cheese Press.   **Right** A cascade in the cave.

YORKSHIRE AND NORTHERN ENGLAND

HIDDEN REALMS

Alum Pot

Alum Pot is a dramatic site. No matter how many times you visit, as you step over the enclosing walls, the place instantly asserts its wonder. There is a sheer drop into an almighty chasm, and this is certainly the most picturesque entrance anywhere in the Yorkshire Dales. The shaft is fed by small streams flowing off Ingleborough and it can be accessed from several points.

Ponder for a few moments upon the bold undertakings of our early predecessors who manfully strode up the hill with a mountain of heavyweight wooden-runged ladders and hemp ropes. Caving pioneer John Birkbeck organised the first serious attempt, via Long Churn Caves, in 1847. One of the team, William Metcalfe, became the first person to reach the bottom of the daylight pitches. The following year, exploration continued for just a few metres to the terminal sump.

The caver of today will drop gracefully down the shaft into this spectacular 104-metre-deep pothole, taking no more than a couple of hours to descend to the bottom and return. Disappointingly, there is little in the way of dry cave to explore. Just a few metres beyond the foot of the last pitch the stream meets the Long Churn water entering from Diccan Pot and then flows over a dark cobbled floor to disappear into the tannin-stained sump pool.

Whether you are a walker or caver, a trip to Alum Pot is a truly magnificent experience.

Left Dolly Tubs traverse.
Right View from the Dolly Tubs ledge.

HIDDEN REALMS

White Scar Cave

YORKSHIRE AND NORTHERN ENGLAND

White Scar Cave is a renowned show cave a couple of kilometres up Chapel-le-Dale from the village of Ingleton. First entered and explored in 1923 by the intrepid Christopher Long, White Scar is a resurgence system that extends back beneath the western flank of Ingleborough to an explored length of six and a half kilometres. The show cave, which was opened in 1925, initially had about 400 metres of passage, but, following the construction of an artificial tunnel in 1991, the massive Battlefield Cavern was incorporated, more than doubling the area accessible to the public.

The Battlefield Cavern is the high point of any visit to this cave. This void is perhaps twenty metres in height, with monumental boulders scattered across every section. At the far end of the enormous fossil passage, undeveloped cave leads on through the Western Front to Long Straw Chamber, which is beautifully adorned with straws up to two metres in length. Here you know you are somewhere very special.

Below the Battlefield Cavern, in the stream passage, lies a deep canal followed by an impressive monolith, the collapsed block – Big Bertha. For the caver this too is memorable. Ahead a boulder choke leads to a spectacular streamway; simply superb with two kilometres or more of spacious passage, cascades and sumps. This is a more committing wet trip, suitable only for dry, stable weather conditions.

Left Long Straw Chamber.
Above The first lake leading to the boulder choke and further reaches.

Gaping Gill

Steeped in history, Gaping Gill – or GG – is undoubtedly Britain's best-known pothole. Situated close to the village of Clapham, in the Yorkshire Dales National Park, this famous pothole received its first exploratory attempt by John Birkbeck in 1842, who reached a ledge at a depth of fifty-eight metres. It was bottomed by the renowned French speleologist Édouard-Alfred Martel in 1895. Some 110 metres below the surface his reward was the largest cavern in the British Isles. This was to be a significant spur to British endeavours in the underground sphere.

For sporting cave enthusiasts the easiest way to view Main Chamber is via Bar Pot, a two-pitch, well-travelled route which remains accessible no matter how much water may be pouring into the ground at any of the other entrances. In total, Gaping Gill has some twelve kilometres of passages leading away from the enormous cavern and the water itself eventually finds its way back to the surface at Ingleborough Cave, which most visitors will have passed on their walk up to GG.

Today, the Bradford Pothole Club and the Craven Pothole Club mount annual 'winch meets' in May and August respectively. At these, members of the public may be transported effortlessly into the depths of the cave in a special bosun's chair to view the incredible spectacle for themselves. It may be a good hour's strenuous walk to reach Gaping Gill, but to be lowered into the huge cavern and view the Fell Beck stream crashing on to the cobble-strewn floor is a sight never to be forgotten.

Above Fell Beck tumbles into the cave entrance. **Right** The main chamber.

Juniper Gulf

To an experienced caver Juniper Gulf is an evocative name that just fires the imagination. This legendary pothole lies on the Allotment slopes of Ingleborough, more than a three-kilometre walk from the nearest parking spot. In the past, accompanied by ladders and lifelines, this was a very serious undertaking. Today, competent SRT practitioners with good rigging can complete the trip in perhaps three or four hours.

Short pitches lead progressively along an ever-deepening fissure and you grasp the character of the place very quickly. Traversing along narrow and increasingly precarious ledges, the sense of exposure is not for the faint-hearted. Aerial acrobatics are required throughout. Rigging and crossing places like the Bad Step are highly memorable, especially when the ledges become greasy and slippery. The sense of wariness grows with virtually every metre.

The ledges remain at the same level, while down below the gurgling water cuts ever deeper. With cowstails clipped and a distinct air of apprehension, you move forward from the foot of the third pitch to the brink of an immense black void. Along with Titan in Derbyshire and nearby Gaping Gill, this final, big, mostly free-hanging, fifty-metre pitch is one of the most fabled shafts in the British Isles.

Juniper Gulf – which has a length of 244 metres and a depth of 128 metres – is an action-packed adventure, an awesome place!

Above left Surface view. **Above right** First section of traversing. **Right** The big pitch from the bottom.

HIDDEN REALMS

Lancaster Hole

Lancaster Hole is located on Casterton Fell and approached via Bull Pot Farm. It presents some of the most spectacular underground scenery in the country. This strategically sited hole was discovered by George Cornes and Bill Taylor on 29 September 1946 and first explored by members of the British Speleological Association.

Today, it forms part of the longest cave system in the British Isles – the Three Counties System – which stretches from Cumbria in the north into Lancashire and North Yorkshire. While the geology and hydrology clearly indicate a veritable web of interconnecting passages, many segments of the overall complex are separated by sumps. Lancaster Hole gives access to the largest continuous section of dry cave in this system with over fifteen kilometres of passages and a depth range of seventy-eight metres.

The entrance is a kilometre of easy walking across the fell. A thirty-four-metre shaft leads directly into a substantial fossil passage and it's only a short distance to find good formations. The finest of these is the Colonnades. Beneath the dry upper levels lies an equally superb streamway, the Lancaster Hole Master Cave, which is one of the keys to the navigationally challenging routes leading to other entrances such as County Pot, Wretched Rabbit and Top Sink. The waterways pose a serious problem in wet conditions, but, once tasted, this is a system that will certainly call you back.

Left The Colonnades. **Above** A swirl hole in the Main Drain.

County Pot

County Pot is a complex labyrinth with over six kilometres of passages, a very popular route into the netherworld under Casterton Fell. Located on the side of the Ease Gill valley, the entrance was excavated by members of the Northern Pennine Club and Red Rose Cave & Pothole Club in 1952. Since then, it has been joined to a number of other entrances and is now subsumed into the ever-expanding Three Counties System.

This is a classic way to enter and explore the vast and complicated easterly part of the Ease Gill network. County Pot has something to suit everyone: short pitches, well-sculpted tunnels and sporting challenges such as the slippery chimney up to Poetic Justice. Many of its passages can be safely visited in most weather conditions. Route finding in the system is not easy; a survey and description are vital. With these in hand, it is possible to undertake extensive through trips, exiting the cave further up or down the valley, or indeed on to a different fell. Along the way, and providing excellent reward for your efforts, are some superbly decorated and nicely accessible places such as Easter Grotto and Gypsum Cavern.

YORKSHIRE AND NORTHERN ENGLAND

Left The second pitch. **Above** Gypsum Cavern.

HIDDEN REALMS

Shuttleworth Pot

Shuttleworth Pot lies on Leck Fell and constitutes a relatively recent addition to the Three Counties System. It was discovered in 1997 by the passing of a sump – 420 metres long and thirty-three metres deep – from Witches Cave. The dry entrance was opened in 2010 by the combined efforts of many northern clubs.

Shuttleworth Pot extends to over one and a half kilometres. With a modest depth of sixty-seven metres and just two pitches to reach the bottom, this is a particularly fine cave that will captivate and intrigue any visitor. Downslope from the second pitch is the House of the Rising Sump with its impressive rock bridges, wild sculpting and metre-high waterfall.

This is an awesome sight. Imagine, as a diver, surfacing from the dark peat-stained water to the stark vista this place presents. Crossing the waterway here will lead you over an ascending boulder slope and gour cascade into the area known as the Exercise Yard.

Upslope from the second pitch leads through a short, low section into Painter's Alley. Beyond is an extensively decorated gallery with a spectacular array of straws and areas of delicate helictites; this is enough of a reward in itself for a visit. Shuttleworth Pot presents a beautiful, relatively easy underground trip that will involve little navigational difficulty.

Above Straws in the large chamber at the head of Painter's Alley.   **Right** The House of the Rising Sump.

HIDDEN REALMS

Notts 2

YORKSHIRE AND NORTHERN ENGLAND

Notts 2 – also known as Committee Pot and the Iron Kiln – is a superb venue and a must-see cave for just about every speleologist. Situated high on Leck Fell, it's reached by a very short walk across the moor.

The entrance itself presents an incredible feat of excavational endeavour, instigated by an amazing advance when the sump at the bottom of Notts Pot was passed in 1985. The divers' discovery was the dream of every explorer – over eight kilometres of large, easy walking passages with sumptuous arrays of calcite formations. Over the following years the location of a possible dry route into the extension was identified and a mega shaft-sinking exercise was undertaken. The diggers achieved their goal in late 2000. In November 2011, Notts 2 was further connected to Lost John's Cave and the Three Counties System was finally established.

Today, the scaffolded entrance shaft drops around twelve metres, there's a short crawl which leads to a further twenty metres of climbable descent, and then you are into open cave. There is so much to see and explore with many inlets feeding the main streamway. Perhaps one of the most notable is Curry Inlet with its fine display of delicate, translucent curtains. Notts 2 is outstandingly beautiful, and this magnificent place will undoubtedly call you back for another visit.

Left Curry Inlet curtains.   **Above** Curry Inlet.

HIDDEN REALMS

Ireby Fell Cavern

Ireby Fell Cavern is one of those great places that everyone should visit at some point in their caving life. It lies off Masongill Fell Lane, about four kilometres north-west of Ingleton, and was first explored in 1949 by members of the British Speleological Association. Any number of important extensions have been made in subsequent years, resulting in a length of over six kilometres and a depth of 112 metres. It is now connected to the Rift Pot and Large Pot to the east, while downstream it connects with Notts Pot via a sump and thus links into the monumental Three Counties System.

A well-constructed concrete pipe furnishes the entry shaft and a steep slope beneath leads directly to the first of three pitches forming the traditional route down. The naming of these is memorably exquisite: Ding, Dong and Bell. Given the immediate proximity of the first two drops one can imagine how they acquired their names. The streamway from Bell leads to the climbable cascade (Pussy), then the fourth drop (Well Pitch), continuing the rhyme. None of these are lengthy, which accounts for the popularity of Ireby for training. Once down, there's a splendid section of fossil passage: Duke Street. This is a fine stretch of shapely rounded passage leading to the first sump and a fitting end to a sporting day underground. But there's a lot more to Ireby Fell Cavern for those hardy enough to take on the challenge.

Above left Looking down Dong Pitch. **Above right** Traverse at the head of Bell Pitch. **Right** Duke Street near Sump 1.

Simpson's Pot

High on the hillside, Simpson's Pot is found on the western side of Kingsdale, opposite Braida Garth farm. It was first explored by members of the British Speleological Association in 1940, but a legendary team from the University of Leeds Speleological Association, including the Brook brothers, achieved a momentous advance in 1966 when they discovered the Kingsdale Master Cave. As a result of their efforts, the Valley Entrance was opened in 1967, establishing one of the finest 'pull through' trips in the British Isles. Today, cavers make a magnificent descent of 112 metres from the Simpson's entrance, down to the Kingsdale Master Cave, exiting through the Valley Entrance.

This is an absolutely delightful cave, although it starts as an uninspiring, but comfortable, crawl. A series of easy, clean-washed pitches and a short, draughty duck take you quickly into the depths. The most memorable obstacle is the entrance to Slit Pot which is notoriously awkward (i.e. impossible!) for larger individuals. Watching someone wriggling through the narrow cleft is entertaining! At twenty-four metres this is the largest drop of the day. (Fortunately, this notorious spot may be bypassed via the awesome Swinsto Great Aven.) One more short pitch and the cave takes on more of a horizontal aspect and soon you are romping down the Kingsdale streamway. The last little hurdle is the 'up' pitch into Roof Tunnel and from there you are virtually out. Simpson's Pot is a thoroughly enjoyable, sporting Yorkshire day.

Left Chandelier Pot. **Top right** Storm Pot leading to the duck. **Right** The duck.

YORKSHIRE AND NORTHERN ENGLAND

HIDDEN REALMS

Rowten Pot

Rowten Pot is without a doubt a northern classic, which was first bottomed by members of the Yorkshire Ramblers' Club in 1897. This is a real chasm of a pothole; it is easy to locate being just to the side of Turbary Road, on the western side of Kingsdale. It may only be 259 metres long and 105 metres deep, but it will provide a very fine technical day for an SRT practitioner.

There are several permutations to reach the bottom, the choice of which will be greatly influenced by the state of the weather. At the outset the options are twofold. Big Gully Route is an atmospheric descent, initially following the stream, which emerges high in the main shaft. The Eyehole Route provides a dry point of entry with an interesting rock bridge and some muddy traverses. These paths converge at the bottom of the impressive main shaft. Below, there are more options – the Direct Route or the Flyover Route – and eventually you land very close to a pair of sumps.

The larger, downstream sump pool connects with Kingsdale Master Cave via three short dives. The smaller, longer, upstream sump leads towards Frakes' Passage, which in turn is another lengthy, diver-only section of the Kingsdale system.

Left The cave is renowned for its spectacular shaft descents. **Above** Big Gully Route entrance. **Right** The dry ledge on the Eyehole Route.

HIDDEN REALMS

Kingsdale Master Cave

The entrance to Kingsdale Master Cave lies in the valley bottom, close to Braida Garth farm. This popular system is the gathering ground for most of the water disappearing on the western side of the valley, taking the flow from Simpson's, Swinsto, Rowten, Bull Pot, Yordas and more.

The Valley Entrance is conveniently sited just twenty or thirty metres from the road; peer over the wall and a well-trodden path leads directly to a quaint circular lid. Slide the cover to one side and slip gracefully into the natural cave just a metre or so beneath the grassy hillside. The fine fossil tunnel you are embarking upon was first discovered in 1966 following a notable exploratory advance at the bottom of Simpson's Pot.

Today, you follow your nose and stoop along a lovely tunnel for a short distance until you meet a six-metre pitch. Below this you land in the waterway, just at the point where the combined flows from up-valley disappear into a foam-strewn sump. This is the start of a cave dive that leads, after 1,800 metres, to Keld Head, the scene of a world record traverse back in 1979. Heading in the opposite direction is a particularly fine stream passage that wends its way through spacious, well-scalloped tunnel for over 300 metres up to the Rowten Sumps. En route you will pass various side leads, the most significant of which is the small waterway which flows in from Swinsto and Simpson's. What a pity the passage doesn't continue further because this is a streamway to savour.

Left Very clear water in the downstream sump.   **Above** Formations in the Roof Tunnel.

HIDDEN REALMS

Illusion Pot

Illusion Pot is a surprisingly impressive site nestling in a shallow doline at the bottom of Kingsdale, on the eastern side of the valley. Were it not that the flooded Dale Barn Cave, in Chapel-le-Dale, was pursued by divers right through, and deep below, Scales Moor, then this superb cave would probably still remain undiscovered. Illusion Pot, therefore, is an amazing extension of the Dale Barn system, taking the overall length of the complex to more than five and a half kilometres.

With an entrance location pinpointed from underground, a masterful excavation project eventually resulted in a twenty-six-metre shaft which today gives dry access into the cave. It's then an easy crawl to reach what the divers referred to as Dale Barn 3; for them this was a particularly committing and remote place. You can well imagine the excitement for the original discoverers as they emerged from the sump to find places such as the beautifully decorated Rushton Chamber and the large, spectacular fossil tunnel now called the Expressway. A thorough inspection will reveal an exciting traverse and other grottoes with a profusion of straws which render the place an exceptional experience.

Today Illusion Pot presents a lesser-known, easy but thoroughly enjoyable caving trip, as well as providing a valuable insight into the complex evolution of the caves of the Yorkshire Dales.

Above left Straw Grotto at the end of the Expressway.
Above right Rushton Chamber. **Right** Sump below the Brunel Rift.

HIDDEN REALMS

Sleets Gill Cave

Sleets Gill is a splendid cave. It has a surprisingly complex hydrology which means that a period of fine, stable weather is recommended before contemplating a visit. The entrance sits high up the side of the valley and when you see the location you would be forgiven for thinking that the opening was perhaps a fossil sink, where water once entered the ground. But the reality is quite the opposite. The steep slope down into the hillside is a flood resurgence, a remarkable place by any standards.

Sleets Gill is now memorable for the highly audacious rescue in March 1992 when Roy Deane and Les Hewitt unexpectedly became trapped by rising water. The event climaxed when it became clear that unless divers were sent in to bring them out, the pair were surely going to drown. Venture here and you will be amazed at the feat the rescuers pulled off. From an easy stoop the entrance slope degenerates to a flat-out crawl over loose rocks. It's interesting to contemplate water flowing *up* this section and then consider transporting two non-divers through here to safety.

The cave suddenly enlarges, and the Main Gallery is a magnificent broad phreatic passage akin to a railway tunnel. Splashing through shallow puddles it's a joy to stroll along. This is a pretty unique place that ends abruptly at a boulder choke after the best part of 400 metres. Just before the blockage you discover the stream which accounts for the flooding. In dry weather a low waterway can be followed further through the aptly named Hydrophobia Passage. The history of the cave will deter any lengthy sightseeing, but visiting this 2,400-metre-long system, near Hawkswick in lower Littondale, provides a very interesting subterranean experience.

Left Calcite boss in the Main Gallery. **Above** Hydrophobia Passage.

Dow Cave

Dow Cave lies a couple of kilometres up-valley from Kettlewell in Wharfedale. It is often memorably associated with Providence Pot (via the Dowber Gill Passage) located two kilometres to the south-east. The combined length of the system extends to 3,700 metres.

Dow Cave is lovely; standing at the entrance of this resurgence network, a very large and impressive canyon wends away into darkness. Some 300 metres of clean-washed, scalloped tunnel later, the low-arched entry to the notorious Dowber Gill Passage feeds in from the right. Passing this, the main route leads onwards crossing pools and cobbled floor to reach a large boulder blockage. This can be passed by a precarious route known as Hobson's Choice, where care is needed.

Beyond lies a different sort of area, one prospected by lead miners. A series of chambers hereabouts will reveal short workings dating back over 150 years, but evidently the site was not as rewarding as they might have hoped. The main cave continues as a good, sporting passage. There are formations en route, including the aptly named stalactite Goliath, and the high-level Roof Gardens. The couple of hours spent looking around Dow Cave will certainly be enjoyable.

Having sampled this fine cave, you will be wise to temper your enthusiasm for tackling the Dowber Gill Passage. The traverse between Providence and Dow is a classic collector's piece. Dowber Gill is a deceptively straight and easy line by survey, but infinitely more challenging, and distinctly exasperating, to negotiate.

Top left The duck at the lower end of Dowber Gill Passage. **Above** Meandering waterway.

Mossdale Caverns

Mossdale is one hell of a cave. Drown or Glory swims, Kneewrecker series and Far Marathon crawls – names that speak volumes and give clear indication as to what may be expected.

The first determined exploration was in 1941 by the legendary hardman Bob Leakey. Mossdale is a gruelling undertaking for the fittest of cavers, and added to its physicality is the fact that in even moderately wet weather the place floods to the roof. Many of Leakey's visits were undertaken alone and this extraordinary man was known to have dug out a trench and slept in the Mud Caverns, far from the entrance.

However, this site is more often recalled for the tragedy which occurred here. On 24 June 1967 a party of leading young cavers was heading to the end of the system aiming to further the exploration. A damp, dismal morning turned into a rainy afternoon. Mossdale Beck swelled and began to pour increasing amounts of water into the entrance and down the constricted passages. For the six men below there was no sanctuary and, despite the tremendous efforts of rescue teams, all lost their lives. Their memory lives on.

Yet there is far more to this system. On a fine day, with a dry forecast, a trip to Rough Chamber is quite an experience: clean-washed, well-sculpted passages and the impressive Boulder Hall. But the greatest allure remains the huge potential for further discoveries. Lying beneath Grassington Moor, the cave is over four kilometres away and about 140 metres higher than the resurgence at Black Keld. Somewhere between, there is assuredly one very big cave system waiting to be found …

YORKSHIRE AND NORTHERN ENGLAND

Left Broadway stream passage.   **Top** Boulder Hall.   **Above** The Swims.

HIDDEN REALMS

Stump Cross Caverns

Above Reindeer Cavern. **Right** The Bowling Alley.

Stump Cross Caverns are well known as a show cave. Located on a bleak, windswept hill midway between Pateley Bridge and Grassington, the total system extends to over 3,300 metres. This place was first discovered by two prospecting lead miners – Mark and William Newbould – in 1860. The pair had moved to the area from Derbyshire and, despite not finding worthwhile lead ore, the miners realised that they had chanced upon a site with evident tourism possibilities. As a consequence, the enlightened entrepreneurs left the formations in the cave intact and adventurous tourists began visiting from the early 1860s.

The show cave utilises the upper levels of the complex and there are some lovely areas of formations, both in and beyond the public section. Bones of reindeer and wolverine have been found so there is also an archaeological interest at the site.

In more recent times, Stump Cross was briefly connected to the neighbouring Mongo Gill System but currently the two caves are separated by a debris blockage. With potential for future links to other nearby sites, this is certainly an interesting area for both cave and mine enthusiasts alike.

Goyden System

The Goyden System lies near the head of Nidderdale, up the valley from Pateley Bridge. It comprises three sections: Manchester Hole, Goyden Pot and the furthermost downstream part of the complex, New Goyden Pot. In total, the network has over six kilometres of passages which, it must be understood, all flood completely.

Goyden has an easy walk-in entrance and descends steeply through boulder-strewn terrain to reach the River Nidd. Downstream, past the Cascades, a huge canyon leads to a sump. At a higher level lies a complex labyrinth of interconnecting passages.

New Goyden Pot presents an impressively large tunnel below beautifully free-hanging, consecutive shafts of fifteen metres and twelve metres at the entrance. It is certainly worth reflecting that water flows *up* these shafts in times of flood. Wander along the majestic river passage and marvel at the sights; the place is huge, the rock sculpting fascinating and waterborne debris will be spied in the highest cracks. The clean-washed streamway clearly shows that a vast flow passes through the system. This place encourages a healthy respect for the forces of nature.

The river eventually disappears into a downstream sump not to reappear until the Nidd Head Risings, below the small village of Lofthouse, a straight-line distance of around two kilometres from the cave entrance. Whether divers will ever be able to achieve this traverse remains to be seen, as the underwater environment is technical and essential dive lines are often shredded in flood.

Top left Top of the First Pitch in New Goyden Pot.
Left Main inlet in New Goyden Pot.
Right Small cascade in Goyden Pot.

HIDDEN REALMS

Ibbeth Peril Cave

Left and above
High-level primary inlet.

Several small caves lie in the beautiful, somewhat remote valley of Dentdale in Cumbria. The gorge-like setting here is quite charming; Ibbeth Peril 1 is found directly alongside a waterfall, above a circular plunge pool. Depending on the weather, the place may be dry, quiet and benign, but, following a short spell of rain, it presents a spectacular change of face. As Ibbeth Peril takes water, caution is the operative word; this is emphasised when you view the substantial piles of driftwood amassed around and above the cave entrance.

Slither inside and weave your way into an enlarging passage and you will be wonderfully rewarded. Less than 100 metres from daylight you reach the vast expanse of the Main Chamber, littered with monumentally massive boulders long detached from the ceiling above. This cavern is very impressive, as are the unusual black, grey and white formations off to the left. The water has long since disappeared beneath the floor to enter a dank and dismal area below, but the primary stream inlet at the furthest point in the cavern will reveal further sights of splendour. This latter passage is a wonderous, shapely, clean-washed phreatic tunnel of comfortable walking height additionally graced with superb formations at the start.

Ibbeth Peril might only be short, but it makes a visit to subterranean Dentdale an unforgettable experience. It's also a brilliant photographic venue.

HIDDEN REALMS

Coniston Copper Mines

Coniston Copper Mines nestle in the eastern shadow of the Old Man massif in the Lake District. They comprise an extensive complex of very interesting, technical and colourful mines which present the possibility of undertaking a superb underground traverse. Here, an upper entrance, Paddy End, lies adjacent to a tarn, Levers Water. The setting is magnificent, and this feeling is duplicated as you carefully descend into the depths of the abandoned workings. Shaft after shaft – three pull-through abseils – can be descended, along with a network of connecting passageways, to emerge 145 metres lower down the hillside and little more than 500 metres distant from Coniston Coppermines youth hostel.

Underground enthusiasts have done a splendid job reopening and stabilising this section to reveal a wonderous insight to our copper mining heritage. The main route through is generally stable and brightly coloured with fabulous areas of blue flowstone either spread in streaky patches over the walls, piled in small tumps on the floor or simply lending a lovely blue clarity to short canals or pools of water. The trip from Paddy End to the Hospital Level, near Miners Bridge, is undoubtedly a classic, one of the finest mine traverses in the British Isles. And if you are keen to return for more exploration, an alternative trip from Fleming's Level to Taylor's Level is also well worth undertaking.

Top left Blue rock. **Above left** Steep rope descent. **Above right** Blue canal. **Right** Mid descent on the first pitch.

HIDDEN REALMS

Nenthead Mines

Nenthead was once a nationally renowned area for lead and silver mining; indeed, this complex is regarded as one of the most intact mining landscapes in the UK. Located forty kilometres south-east of Carlisle, it is a mecca for mine enthusiasts as many kilometres of tunnels can be visited.

Most of the remains date from the eighteenth and nineteenth centuries, but mining was intensive from about 1690 when the Rampgill vein was discovered. Lead production peaked in the 1820s and was replaced by zinc production later in the century, until mining ceased in 1920. Some of the spoil heaps were then reprocessed for zinc and fluorspar, which continued until 1971. Today, there is a heritage centre displaying the rich history of the area.

There are any number of entrances in the Nenthead complex. Carrs Mine is a good starting point as this could be classified as a show mine, with open days for the public. Rampgill Horse Level is a major site which intersects the highly impressive, hundred-metre-deep Brewery Shaft. Further up the valley lies Smallcleugh Level, another extensive section of mine. Here are found all manner of interesting features and some meticulously crafted stone architecture lining the tunnels.

From the wonderful waterwheel, mine machinery, artefacts, inscriptions and the skilled, beautiful stonework underground, to all manner of industrial archaeological remains on the surface, there is so much to see here that a single visit will never do the area justice.

Left Brewery Shaft waterwheel. **Top left** Calcite formations on rhizomorphs in Rampgill Horse Level. **Above left** Brewery Shaft. **Above right** Skilled stone working in the mine.

SCOTLAND

Scotland is not famed for its long or deep caves, but, for those who seek them out, there are beautiful and sporting places to visit.

Applecross lies on the scenic west coast. Here, hidden within a small valley are wonderfully decorated caves with some unusual geological features. Gems of the underworld.

The very remoteness of the caves of Assynt is special. Sites are located in windswept valleys with dry riverbeds and alpine flowers. Here the underworld complements your experience with rushing waterways, climbs, traverses and significant archaeological interest.

Even further north is the charming tourist attraction of Smoo. This is a bustling and busy coastal venue, a true contrast to the wilder sites of Scotland.

84 Applecross Caves
85 Uamh an Claonaite and Rana Hole
86 Smoo Cave

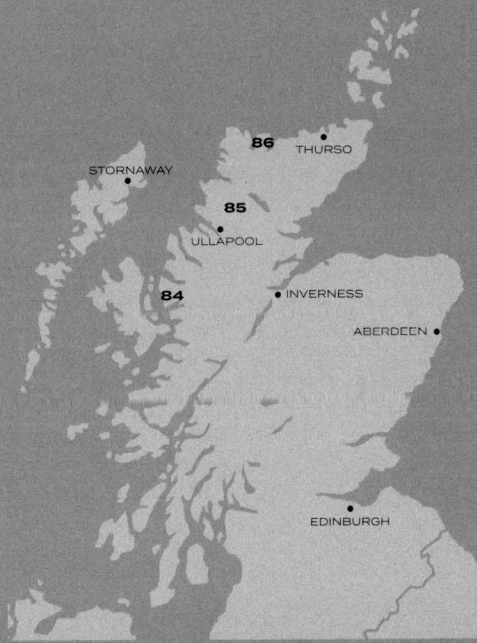

Left Rana Hole entrance, Assynt, Sutherland.

HIDDEN REALMS

Applecross Caves

Applecross Caves lie on a remote and rugged peninsula in the Western Highlands – directly opposite Raasay, Rona and Skye. Located a kilometre or so from the small community of Applecross, this network of caves is largely unknown and unvisited, even among the most dedicated Scottish cavers.

It's only a short walk to the quiet valley where finding the entrances, concealed by undergrowth, might prove a little challenging. While there are a number of caves, the reality is that they are all part of one system. From Poll Breugair (Liar's Sink) at the upper end of the valley to the flood resurgences at the bottom end of the network, it is interesting to speculate upon further discoveries and a complete through trip at some point.

Water levels are crucially important for access as some sections become impassable in bad weather. This is certainly the case in Uamh nam Fior Iongantais (Cave of True Wonders).

This flood resurgence is the real jewel in the crown where in 2011 the finest grotto in the area, indeed the whole of Scotland, was revealed. Another spectacular sight is the unusual geological formation Canada Cavern in Poll Breugair. With its fault running along one wall and domed roof, this place is impressive.

Applecross Caves may be a recent addition to the underworld of Scotland, but any caver will be pleasantly surprised by the underground sights in this complex.

Above Grotto in Poll Breugair (Liar's Sink).
Right Canada Cavern. **Opposite** Uamh nam Fior Iongantais (Cave of True Wonders).

HIDDEN REALMS

Uamh an Claonaite and Rana Hole

The remote site Uamh an Claonaite is Scotland's most significant cave system. It lies in the parish of Assynt, on the west coast of Sutherland. There are currently over 3,400 metres of passages, with a depth of 110 metres. The complex has two entry points and today it is only a single sump that prevents a complete non-diving traverse from one to the other.

The first entrance, Uamh an Claonaite, was located in 1966 and significant discoveries resulted after diving breakthroughs in 1975 and 1995. Entry to Claonaite, or the 'Old Cave', is via a jumble of loose boulders but thereafter the streamway is what this lovely system is all about. The water tumbles away over first one and then another set of cascades. It's just a great sporting descent. A scramble here, a traverse there, foaming plunge pools, a couple of sump bypasses, nice clean passage; apart from dripstone deposits this cave has it all.

The second entrance – Rana Hole – required a prolonged excavation which finally gave 'dry' access to the furthermost reaches, beyond the sumps at the end of Claonaite, in 2007. The descent is dry and begins as a series of vertical drops using ladders and SRT. Along the way, cavers negotiate an interesting descending traverse involving fixed metal aids which eventually leads down into the horizontal part of the cave. Here there are immense fossil passages, not least the Great Northern Time Machine, and exceptional archaeological finds have been made. This is a very different experience to that of Claonaite.

Above The vast expanse of the Great Northern Time Machine. **Right** The Falls of Jabaroo in low water conditions.

HIDDEN REALMS

Smoo Cave

Above Entrance area. **Opposite left** Smoo Cave. **Opposite right** Swimming in the cave.

SCOTLAND

Smoo Cave is a well-known tourist destination in the extreme north of Scotland. Located just over a kilometre and a half from the town of Durness, in Sutherland, this place is impressive. The name Smoo comes from the Norse word 'smuga', meaning cave. Archaeological studies have revealed numerous Viking remains and found clear evidence of ship building and repair.

It's only a few metres from the parking area down to the beach to view the entrance, which is freely accessible at all times. The first recorded visit was in 1760 but the cave's tourist status only came to the fore after the famous novelist Sir Walter Scott published an account of his tour of the area in 1814.

In reality, Smoo is a cave of two parts. The entrance portal is a large sea cave; beyond this, a narrowing leads over a wooden walkway to a slightly smaller chamber evidently carved by normal cave forming processes. This second cavern is occupied by a deep-water lake, scoured clean by an intermittent surface stream which cascades down from the limestone sink above. In summer, tours are conducted by inflatable dinghy across the watery expanse into a section of walking passage on the far side. In very wet weather, the waterfall thunders in making the lake and cavern a hostile place.

I would recommend Smoo to anyone visiting this part of Scotland, perhaps when travelling the North Coast 500.

IRELAND

The Emerald Isle ... famed for its rugged coastline, folk music and the finest pint of Guinness. Yet there is so much limestone in Ireland and caves are found in almost every county. The Burren, which is centred in County Clare in the Republic of Ireland, is home to some spectacular, but flood-prone, river caves. Tall, sinuous canyons, cascading inlets and decorated grottoes characterise this area. In County Fermanagh in Northern Ireland there is such variety – altogether more substantial caverns, fine pitches, vast galleries and splendid arrays of formations. Many other impressive sites are located far and wide in places such as Kerry, Mayo and Tipperary.

While caves are undoubtedly the principal attraction, Irish mines should not be discounted. In particular, the Silvermines area, where Shallee Mine is situated, is utterly magnificent.

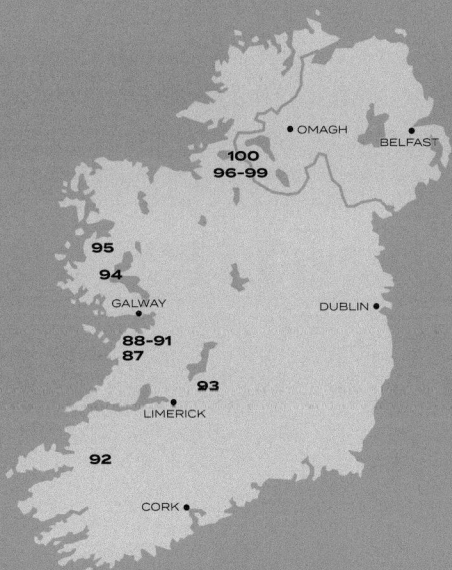

- **87** Doolin River Cave
- **88** Faunarooska Cave
- **89** Polldubh Cave
- **90** Poulnagollum–Poulelva Cave System
- **91** Cullaun 2
- **92** Crag Cave
- **93** Shallee Mine
- **94** Ballymaglancy Cave
- **95** Aille River Cave
- **96** Marble Arch Caves
- **97** Pollnagollum of the Boats
- **98** Shannon Cave
- **99** White Fathers' Caves
- **100** Noon's Hole and the Arch Cave System

Left View across the Burren, County Clare.

HIDDEN REALMS

Doolin River Cave

A traverse of Doolin River Cave is an absolutely classic trip, probably the most popular of any in Ireland. It presents lots of variety and little technical difficulty. The gently dipping passages drain water from the area south-west of Lisdoonvarna to the coast near Doolin.

The intricate system runs to over ten kilometres and much of the cave lies only just below the surface. Cavers generally make their journey from one of two starting points – Aran View Swallet or the more popular St Catherine's 1. Either way, they will end at Fisherstreet Pot, a thirteen-metre, sheer-sided hole from the surface which will have been rigged before heading underground!

St Catherine's 1 starts as a low, wet crawl. The route quietly gets bigger until you are romping through a classic, sinuous stream passage, so characteristic of County Clare. There are some lovely grottoes en route and you'll pass the memorable Aille Cascade, where an inlet from the surface river pours in from high on the right-hand side. Gradually the canyon lowers to bedding cave as you approach the exit at Fisherstreet Pot.

Plan your day wisely, as this cave can take in a lot of water in wet weather and, while there are plenty of places to take sanctuary, you wouldn't want to get down to the bottom end and not be able to get out. If you've been even more thoughtful, you'll also have some money in your bag with which to walk the few metres to one of the village bars and celebrate your successful trip in a time-honoured manner.

Left Clean-washed canyon. **Centre** Aille Cascade. **Right** Fisherstreet Pot.

HIDDEN REALMS

Faunarooska Cave

IRELAND

Faunarooska Cave is located on the upper slopes of Slieve Elva and is one of the more popular sites in the area. Descending a short passage to a T-junction you pick up a small stream and follow this along a lovely smooth-walled, meandering canyon passage. Yes, there may be a fair amount of sideways shuffling, but there's never anything desperate. The twists and turns, together with short, gentle water chutes, are interesting and leisurely. After the first and second cascades, where the stream drops more steeply, the nature of the place subtly changes. The roof lowers and there's a lot more flowstone. At a small waterfall the cave suddenly transforms again, this time into a phreatic tunnel. Just when you feel that the promise of larger passage beckons, the stream disappears as it plummets and splashes its way down a fractured side development into a tight sump.

Continuing directly ahead takes you into an old, long-abandoned fossil passage. A roof traverse leads to a well-decorated section and ultimately to a superb twenty-two-metre pitch. The cave now delivers another cruel twist, for no sooner do you reach the bottom than you arrive, almost immediately, at a second (static) sump, the end of this 1,690-metre system. Faunarooska is a particularly fine and enjoyable day out and really sporting in wet weather.

Left Curtains in the stream passage.   **Above** Grotto near the end of the streamway.

HIDDEN REALMS

Polldubh Cave

The caves of County Clare are noted for their flood-prone nature and none more so than those in the Coolagh Valley. But while Coolagh River Cave itself must be treated with extreme caution and respect, up in the headwaters of the system's catchment lies the ever-popular and forgiving network of Polldubh. This system, draining the southern Knockauns Mountain and south-west Slieve Elva, presents a great venue when everything else is impassable. Yes, it takes water, and the place will be wet, but there's something for everyone here.

Polldubh has several entrances and two principal streamways, offering the possibility of an easy crossover trip. Shaley rocks litter the floor throughout rendering footwork somewhat slippery, but even so it's a good place for novices. The three-metre-high cascade near the confluence of the two branches of the cave will add a touch of drama in wet weather, but won't pose a threat in any conditions. This is an easy cave and if you look closely there are some fine formations. You will probably not wish to explore the terminal downstream extremities as these are low and grovelly, but otherwise this place provides a pleasant subterranean excursion. The cave has a total length of over 1,400 metres and there is minimal requirement for vertical equipment.

Below Beautiful formations.
Right Cascade at the confluence.

HIDDEN REALMS

Poulnagollum–Poulelva Cave System

Poulnagollum–Poulelva is the longest cave system in Ireland, extending to over sixteen kilometres with an estimated depth of around 100 metres. It lies on the north-east slope of Slieve Elva, six kilometres from Lisdoonvarna in County Clare.

The Poulnagollum entrance is a short and steep rock wall, best rigged with a handline due to its slippery nature. You soon find yourself heading down into glistening clean-washed passage – everything about this system shouts 'water course'. The rock is well sculpted, and scallops of every size are etched into the surfaces. It doesn't take much imagination to appreciate that when it rains a lot of water passes through this place on its way to the Killeany Resurgence, three kilometres distant.

Given the overall extent of the complex, some fine excursions are possible; in particular, through trips to or from Poulelva Pot, 1,500 metres to the south. The sheer drop at the latter site is impressive, secreted in its jungle-like setting. Like any big system, route finding can be a little challenging, but this is certainly a cave that will call you back for longer ventures such as the superb circuit incorporating the Branch Passage Gallery.

Left Formations in the Main Streamway.   **Above** The main Poulnagollum entrance.

Cullaun 2

This is a lovely cave and is capable of being enjoyed in almost all weather conditions. It offers a fair variety of terrain with the possibility of making a satisfying circuit or two involving, for example, the Old Streamway or the Year Passage route. The initial section is a wonderfully sinuous dry rift passage, which emerges at the Cascades, the beginning of the Main Streamway. From here on, the canyon-like continuation is easy-going. There are formations – including straws and helictites – in the roof, but the most notable landmark is the calcited boulder partially obstructing the passage, affectionately known as the Bloody Guts. Waist-deep water at Pool Chamber is hopefully the wettest you will get and the point where a high-level fossil passage can be followed back to the Cascades. Continuing downstream leads to a ten-metre pitch, the only technical part of the trip, and quickly after this the 'end' of the cave is reached at a sump.

Cullaun 2 lies on western Poulacapple hill approximately seven kilometres from Lisdoonvarna. It is around 3,500 metres in length and is the most popular in the Cullaun group of caves. The site, like most in the Burren region, was first explored by members of the University of Bristol Speleological Society, in 1952.

Above left Squeeze in Upper Dry Route.   **Top right** The Bloody Guts.
Above right Stream canyon passage.   **Right** Traverse beneath straws.

HIDDEN REALMS

Crag Cave

Until 1983 Crag Cave was just 50 metres in length and enticed few visitors. A small stream flowed sluggishly into a fair-sized chamber, and it appeared that the prospects for more passage were poor. It looked as though any continuation, if there was one, would be flooded.

A surprisingly short, ten-metre dive that Easter surfaced into a most spectacular, beautifully decorated passage. Very soon the superb extension was accessible to non-divers and over the following years the near environs were developed as a tourist attraction. Situated two kilometres from Castleisland in County Kerry, Crag Cave is now arguably Ireland's finest show cave.

Beyond the tourist route lie a further three kilometres of cave. Large mudbanks and stream passage give way to the stalactite-encrusted Fangorn Forest, while in a side grotto the White Tower stands in splendour amidst its own attendant draperies.

This is what cave exploration is all about! To stand on the edge of the unknown is an incredible privilege. To be the first to look upon such wonders is fantastic and to share some of those sights with the world at large is so immensely satisfying.

Above John Cooper: the breakthrough in 1983.   **Right** Straws and stalagmites in Fangorn Forest.

HIDDEN REALMS

Shallee Mine

Above The Cathedral. **Opposite** Ore chute.

Silvermines, County Tipperary, is an area rich in minerals and highly significant in Irish mining history. The extraction of metals hereabouts can be dated back to 1289 when Italian miners from Genoa and Florence came in search of silver. Commercial quantities of lead, copper, zinc, barytes and sulphur have been extracted over time but activity was frequently stop–start.

Three kilometres west of Silvermines village, Shallee Mine was worked for copper and lead from around 1770 then intensively for lead in the 1950s. Operations across the Silvermines region had ceased by the early 1990s; today the whole area is silent, the landscape largely reclaimed by nature.

Shallee Mine is a true wonder of subterranean Ireland; nowhere is it more amazing than at the area named the Cathedral. A huge entrance leads directly into a spectacular, partially flooded cavern, the roof of which is supported by massive stone pillars. Seeing this place with sunlight streaming down on to the golden-coloured walls is an unforgettable sight. The water here is deep and gives a clear indication that many more workings lie below.

Follow a side tunnel and access is gained to far more of the mine. It will shortly reveal perhaps the largest underground caverns in Ireland, if not the British Isles. Vast, echoing cavities, a flooded incline with *in-situ* rails, and well-preserved ore chutes all add substance to any visit. It is an incredibly impressive place!

HIDDEN REALMS

Ballymaglancy Cave

Remote from the principal caving areas, Ballymaglancy Cave is located between Clonbur and Cong in County Galway. It may only carry a small stream, but what happens to this water is fascinating. Following a subterranean route, the trickle from Ballymaglancy joins the legendary volume of water swallowed underground in and around Lough Mask. The combined flow then makes its way for some 1,500 metres to reappear at the spectacularly large risings at Cong. No one who has seen them would forget this very impressive sight.

Ballymaglancy has been known for many years and in bygone times was a modest tourist attraction. Today, it is popular with outdoor groups and offers the explorer a 530-metre-long through trip. You enter at a low opening and very quickly an absolutely delightful cave opens before you. Barely inside, a display of fossilised coral gilds the floor and, a short distance beyond, the stream plunges over a cascade. This three-metre drop can be easily passed by traversing along a ledge on the right-hand wall. Thereafter the passage meanders gently, leading eventually to an area where the sluggish flow wends between mud banks skirting the walls. There are some lovely formations in this furthermost section then, all too soon, you reach the small and unimpressive terminal sump. But look to the right and a short, dry passage will lead you out into daylight.

This is an easy cave and a lovely trip in a hydrologically interesting area.

IRELAND

Left The large main passage. **Above** Gours in the main passage.

HIDDEN REALMS

Aille River Cave

IRELAND

Aille River Cave is an awe-inspiring site situated eight kilometres from Westport in County Mayo. Even in dry weather a substantial stream enters the massive crag-lined sink. This has an immense catchment area as is evidenced by the silt and flood debris piled high all around. The cave was first explored by members of the Craven Pothole Club in 1967 and 1968, although this site is reputed to have witnessed an incarceration of local people during the darker days of Oliver Cromwell.

Access to the entrance is via a steep climb down to a large tunnel with a fast-flowing, tannin-stained waterway. Heading downstream, the walls are black and the cave exudes a sense of deep foreboding. The place seems to swallow your light and real care needs to be exercised as you negotiate sections of sharp, brittle and crumbling rock. The shapely, chiselled passages have impressive dimensions, but shredded ropes bear silent testimony to the fluctuating water levels and evident power of the subterranean river. This is a dramatic place.

Deepening water leads on and on until it reaches the final sump. The river is not seen again until it reappears over three kilometres away at the Bellaburke Resurgence which has been dived to a depth of 113 metres, the deepest cave diving site anywhere in the British Isles.

Left Traverse over deep, fast-flowing water.   **Above left** Dinghy on deep water.
Top right Lithostrotion fossil bed.   **Above right** Dry tunnel, a rare feature in this cave.

HIDDEN REALMS

Marble Arch Caves

Marble Arch Caves is a site with a very rich history. The area at the foot of Cuilcagh mountain, County Fermanagh, was a celebrated tourist spot from the 1700s. Édouard-Alfred Martel and Dublin naturalist Lyster Jameson visited in 1895, Martel suggesting this impressive site had potential to become a show cave.

In 1908 the Yorkshire Ramblers' Club commenced activities and over the ensuing years all manner of clubs and individuals have contributed to the speleology of the area. An extensive complex of caves and several different waterways converge at this point. Gradually, passable connections have been made, both above and below water, joining Marble Arch to Cascades Rising, Cradle Hole and other nearby caves. Today the system extends to over eleven kilometres in length, the second longest on the island of Ireland.

In the 1980s, sections of Marble Arch were developed as a show cave by Fermanagh District Council and visitors may be transported by boat via the Wet Entrance before following a path through the large, decorated passages of Skreen Hill 1. Cavers are free to explore exceptionally big river chambers and lovely formations beyond this point or to venture on more challenging through trips to other entrances. Be in no doubt, this is a grand system and Marble Arch Caves present a wonderful experience for tourists and cavers alike.

Above left Crystal-clear water in the Swann's Way sump. **Above right** Legnabrocky Way.
Right Grand Gallery.

HIDDEN REALMS

Pollnagollum of the Boats

Pollnagollum of the Boats is one of the most notable caves in County Fermanagh. The water flowing through this site – the Owenbrean River – drains from the northern slopes of Cuilcagh mountain via the Pollasumera sink and, having passed through an impenetrable choke, eventually discharges into the major system of Marble Arch at the head of the Skreen Hill Passage.

Left Deep water crossing.
Above Large stalactites.

Pollnagollum lies a mere few metres from the Marlbank Scenic Loop road; an easy, steep descent leads, via a substantial boulder choke, directly to a spacious deep-water canal. This may be traversed by a thirty-metre swim but taking a dinghy feels altogether more of an adventure! The reward is an amazingly large chamber immediately beyond which is one of the most splendid stalactite displays to be seen in the whole of the island of Ireland. Travelling to just this point will be highly memorable but the cave continues to follow the river for a further 500 metres, sometimes wading, passing more attractive formations. Pollnagollum is not accessible in high water but under normal weather conditions this is a cave that deserves to be set high on any underground itinerary.

HIDDEN REALMS

Shannon Cave

Shannon Cave is the third longest cave in Ireland and one with immense future potential. The site straddles the border between Fermanagh and Cavan in the western drainage area of Cuilcagh mountain and is approached via the Marlbank Scenic Loop.

The discovery of Shannon Cave fell to members of the Reyfad Group in 1980 but the dangerously unstable entrance series collapsed and sealed the cave some fifteen years later. In 2005 a new point of entry was found via Polltullyard. Since access was first gained here, members of the Irish Caving Club have tenaciously pushed forward the exploration.

Today, Shannon Cave is a joy to visit. The easy, safe entrance of Polltullyard gives access to twenty metres of walking and then a magnificent thirty-three-metre free-hanging pitch. Beyond, some low, mixed passage leads via the entertaining squirm, Rebirth Canal, into JCP Passage in Shannon Cave. From here a once enormous tunnel (festooned with tumbled boulders on one side) leads down a streamway – the underground source of the River Shannon – and on into the depths of the cave.

For the fit and ambitious, boulders, loose chokes and some spectacularly large tunnel will lead ultimately to the furthermost reaches. But many will linger in JCP Passage to marvel at the wonderful arrays of formations along its walls. This is a superb caving trip.

Above Straw grotto in JCP Passage.
Right Exiting the Rebirth Canal.
Opposite Looking up the thirty-three-metre pitch.

HIDDEN REALMS

White Fathers' Caves

The streams that rise in north-west Cuilcagh unite and flow for several kilometres to Lough MacNean. Along the way, near the border village of Blacklion, the stream passes through a set of three shallow caves, known collectively as White Fathers' Caves or St Augustine's Caves. The caves may not be significant in terms of length or depth, but they are without doubt fascinating, unusual and colourful.

White Fathers' is easy to access, and visitors will pass quickly through the first two sections. Just inside the final cave the roof is pockmarked with holes that admit daylight and trailing tendrils from the plants above. Leisurely walking takes you on. The place is wonderfully clean washed and, where not lavishly draped with flowstone, it exhibits superb rock sculpting. The waterway presents canals, bubbling cascades and a small waterfall. The final exit entails deep-water wading or swimming and it should be noted that after heavy rain these lower reaches will sump. Small amounts of flood debris can be observed high up on the walls, but what will impress itself on everyone is the amazing profusion of formations throughout.

This is a splendid little cave and offers so many photographic opportunities ... a wonderful place that every caver should visit at some point. This interesting site is adjacent to a large property that once belonged to the White Fathers (Missionaries of Africa), a Roman Catholic missionary society. Today it is utilised as a prison.

Left Little water spout. **Above** Rainbow Chamber.

HIDDEN REALMS

Noon's Hole and the Arch Cave System

Noon's Hole is a spectacular shaft on Tullybrack mountain in County Fermanagh. At its foot it gives access to an extensive system of passages: one heading further into the hillside and another out to the Arch Cave Resurgence. The overall length of the system exceeds 3,500 metres and has a depth of 108 metres.

Noon's has a colourful history. 1826 saw the unfortunate demise of Dominic Noone, cast into the shaft for betraying his nationalist comrades to the authorities. In 1895, Édouard-Alfred Martel ventured a short distance into Arch Cave and made a partial descent of Noon's Hole. At the latter site in 1912 members of the Yorkshire Ramblers' Club reached a sump, ninety-two metres below the surface. A connection from Arch Cave to Noon's Hole, via Noon's Sump, was finally achieved in 1973.

Left Duncan's Duck, Arch Cave.
Above Noon's Hole.

Today, there is a dry bypass to Noon's Sump and a classic trip will descend the entrance shaft, take you through Crucifixion Crawl and into High Noon's, a varied passage with columns, pools and a large boulder choke. Beyond is the awkward connection into the large, speleothem-rich streamway of Arch 2, which ends well over a kilometre later at a downstream sump. A way past this sump has eluded cavers but, for cave divers, the complete through trip from Noon's to the Arch Cave Resurgence is one of the very best, most sporting traverses in the British Isles.

Glossary

Adit An opening driven horizontally into the side of a hill to provide drainage or access to a mineral deposit in a mine.

Bedding cave A cave, or part of a cave, formed between layers of sedimentary rock, often having characteristic wide, low-roofed passages.

Boulder choke A pile of rocks or boulders, as from a collapsed roof, often blocking further passages.

Calcite Calcium carbonate, derived from the limestone bedrock, deposited in the form of a speleothem such as stalactites and stalagmites.

Choke See *Boulder choke*.

Crossover trip Where two groups enter a system at different points, and each make their way to the other entrance. Often undertaken in vertical systems allowing each group to exit on the ropes rigged and descended by the other team.

Curtain A thin, fluted sheet or draping of calcite resembling a curtain.

Dig An excavation above or below ground to find a new cave/mine or section of cave/mine.

Doline Surface hollow caused by solution of limestone or the collapse of an underground cave.

Duck A section of passageway almost completely full of water. There is normally a small amount of airspace requiring the caver to partially duck their head under water to pass the obstacle.

Exchange trip See *Crossover trip*.

Flowstone A continuous sheet of calcite, covering a cave wall or floor.

Fossil passage A passage abandoned by the water that formed it.

Free dive The passing of a short, flooded section of passage, by holding one's breath, not requiring the use of specialist diving equipment.

Hand-picked Use of hand tools alone to extract mineral ore or to cut a passage in a mine.

Helictite A speleothem displaying erratic or eccentric growth.

Hydrology The scientific study of water and its movement.

Hydrothermal The action or effect of heated water on the earth's crust.

Kibble A large bucket-like container used to haul ore up a shaft.

Level A horizontal tunnel or roadway giving access to the workings of a mine.

Ore chute An inclined passage, about one square metre in size, for the transfer of ore to a lower level of a mine.

Phreatic A cave or passage formed below the water table.

Pitch A vertical section, usually requiring the use of a ladder or rope.

Pot A vertical shaft, either open to the sky or inside a cave.

Resurgence The point at which underground water emerges at the surface, i.e. a spring.

Rift A cave passage that is relatively high and narrow.

Rig To arrange ropes and equipment enabling safe negotiation of a traverse, ascent or descent.

Scallops Water-worn hollows in a cave wall, floor or roof.

Single-rope technique System of exploring caves and mines using single ropes on pitches to enable abseiling and prusiking.

Sink A place where surface water disappears underground.

Slocker See *Swallet*.

Speleology The exploration and scientific study of caves.

Speleothem Any cave or mine deposit or formation, e.g. stalactites, stalagmites and helictites.

SRT See *Single-rope technique*.

Stalactite A formation, usually of calcite, hanging from above.

Stalagmite A formation projecting upwards from the floor.

Stope An excavation in a mine from which ore is, or was, extracted.

Straw A hollow, thin-walled tubular stalactite.

Sump A point in a cave or mine where water completely fills the passage. Short sumps may be free-dived; longer sumps can be negotiated with the aid of specialist diving equipment.

Survey A map of all or part of a cave or mine system.

Swallet An opening in limestone where a stream vanishes underground.

Through trip A trip where entry and exit are from different entrances.

Traverse
1 A passage which is followed by travelling high above the floor.
2 To travel through a cave or mine.

Tyrolean crossing A method of crossing through free space, between two points, attached to a taut rope or wire.

Vadose A cave or passage formed by water flowing under normal atmospheric pressure.